Hinged Dissections

Swinging & Twisting

If you enjoy beautiful geometry and relish the challenge and excitement of something new, then the mathematical art of hinged dissections is for you. Using this book, you can explore ways to create hinged collections of pieces that swing together to form a figure. Swing them another way and then, like magic, they form another figure!

The profuse illustrations and lively text will show you how to find a wealth of hinged dissections for all kinds of polygons, stars, crosses, and curved and even three-dimensional figures. For an added challenge, you can try using different kinds of hinges for twisting and flipping pieces. The author includes careful explanation of ingenious new techniques, as well as puzzles and solutions for readers of all mathematical levels. If you remember any high-school geometry, you are already on your way.

These novel and original dissections will be a gold mine for math puzzle enthusiasts, for math educators in search of enrichment topics, and for anyone who loves to see beautiful objects in motion.

Greg N. Frederickson is Professor of Computer Science at Purdue University. After teaching mathematics in the Baltimore public school system, Dr. Frederickson earned his Ph.D. in computer science from the University of Maryland, and he has been a faculty member both at Pennsylvania State University and Purdue. He teaches and does research in the design and analysis of algorithms and has been funded by awards from the National Science Foundation and the Office of Naval Research. His research papers have been published extensively in computer science journals and he currently serves on the board of *Algorithmica* and the *Chicago Journal of Theoretical Computer Science*. His first book, *Dissections: Plane & Fancy,* was published by Cambridge University Press in 1997.

T0224883

Hinged Dissections

Swinging & Twisting

GREG N. FREDERICKSON
Purdue University

CAMBRIDGE
UNIVERSITY PRESS

CAMBRIDGE UNIVERSITY PRESS
Cambridge, New York, Melbourne, Madrid, Cape Town, Singapore,
São Paulo, Delhi, Dubai, Tokyo, Mexico City

Cambridge University Press
The Edinburgh Building, Cambridge CB2 8RU, UK

Published in the United States of America by Cambridge University Press, New York

www.cambridge.org
Information on this title: www.cambridge.org/9780521010788

First published 2002
First paperback printing 2010

A catalogue record for this publication is available from the British Library

Library of Congress Cataloguing in Publication data
Frederickson, Greg N. (Greg Norman), 1947–
Hinged dissections : swinging & twisting / Greg N. Frederickson.
p.   cm.
Includes bibliographical references and index.
ISBN 0-521-81192-9
1. Geometric dissections.   I. Title.

QA95.F69   2002
793.7′4–dc21   2001043450

ISBN 978-0-521-81192-7 Hardback
ISBN 978-0-521-01078-8 Paperback

To Margaret and Howard Frederickson,
    who came of age in the era of swing,
    and raised a young son who loved to swing!

Contents

CONTENTS

Preface

As readers of my first book, *Dissections: Plane & Fancy,* can now see, I have been at it again. Why? I could blame it on Erich Friedman, who wrote: "loved your book. write another one!" Or even on Don Knuth, who, months before the book appeared in bookstores, related Elwyn Berlekamp's observation that there are three kinds of people in the world: those who have written no book, those who have written one book, and those who have written more than one book. Don then paused and stared at me.

The awful truth is that I could not help myself. At first I knew of so few hinged dissections that I was drawn irresistibly to the task of finding enough new ones to fill a chapter. Then I became intrigued with adapting and discovering dissection techniques that would produce hingeable dissections. And just when I was beginning to feel complacent, Anton Hanegraaf improved on a number of my dissections, which spurred me to redouble my efforts. Finally I recognized the necessity of breaking free from my original style while still engaging readers on several levels. As I wrestled with these challenges, the initial audacity of this project metamorphosed into a cheerful inevitability.

Once again, the intended audience is anyone who has had a course in high-school geometry and thought that regular hexagons were rather pretty. Fair enough, but why should anyone buy a second book on geometric dissections, especially when the requirement that the dissections be hingeable makes finding them considerably more difficult? Isn't one book enough?

Not if you enjoy extraordinarily beautiful objects and relish the challenge of something new. Swinging pieces around on hinges is fundamentally different from just rearranging them. Furthermore, there is a certain tidiness associated with hinged dissections. Gone is that pile of seemingly odd-shaped pieces that demand too much patience to assemble and are ruined if one piece gets lost. Instead, the pieces are all connected, with the connections providing hints for the rearrangement. And for people on a budget who still like to manipulate objects, all you need, in addition to paper and scissors, is some thread and tape. I have tried to make this book self-contained, so you don't need to read the other one first.

There is a minor caveat to this claim of being self-contained: Although I may mention the best known unhingeable dissections by way of comparison, I will generally not show pictures of them. That would make this book twice as long, and besides, those unhingeable dissections are almost all in my previous book.

I would like to thank Anton Hanegraaf for his generosity in allowing me to reproduce so many of his wonderful dissections. But Anton provided more than just raw material. He pushed me beyond my initial efforts on a number of dissections. And he was an unflagging enthusiast of this project from the first day that I wrote to him about it. I would also like to thank Gavin Theobald, Bernard Lemaire, Robert Reid, and Alfred Varsady for permission to reproduce their dissections.

I am indebted to Martin Gardner for sending me Thomas H. O'Beirne's work on bracing various regular polygons. I would like to thank several people who helped keep me going: Bernard Lemaire for his enthusiasm and his helpful comments on the manuscript, Walt and Chris Hoppe for making so many nice models on their laser cutting machine, and Alain Rousseau for programming wonderful animations on the Cabri system. Most of all, I would like to thank Wayne Daniel, who crafted marvelous wooden models and designed and refined the niftiest of hinge mechanisms for them.

I would like to thank Will Shortz for supplying citations to earlier versions of Sam Loyd's puzzles. I would like to thank David Singmaster for sharing with me an electronic copy of his work on sources of recreational mathematics and also xerox copies of Dudeney's puzzle columns in the *Weekly Dispatch.* I would like to thank Alpay Özdural for sharing his work on the anonymous Persian manuscript, *Interlocks of Similar or Complementary Figures.* I would like to thank Sonia Fahmy and Freydoon Shahidi for help in translating a passage from a manuscript. I would like to thank Kathy Trinkle for her assistance with LATEX and font packages.

I would like to thank Robert B. Banks, Erik Demaine, Odette De Meulemeester, Elaine Ellison, Anton Hanegraaf, Volker Priebe, Ron Resch, Dick Ruth, Thomas Siegmund, Sue Whitesides, Herbert Wills, and Ron Wohl for bringing references to my attention. I would like to thank Ernst Lurker and Tim Rowett for sending me commercial models of the twist-hingeable heart. I would like to thank Al Chiscon, Harriet Stone Evans, Elisabeth Hambrusch, Susanne Hambrusch, Anton Hanegraaf, Thomas Horine, and Robert Reid for helping me to improve the text with corrections and suggestions. An extra round of thanks to Thomas and Anton, who were especially thorough in their comments. I would like to thank the University of Illinois libraries, the British Library, the British Newspaper Library, the University of Cambridge Library, the Family Records Centre in London, the East Sussex Records Office, and the Hove Library in East Sussex, all of which I have visited. I would like to thank the Bibliothèque nationale de France for making copies of two manuscripts available to me. I would also like to thank the library staff at Purdue who handled my multitude of interlibrary loan requests, as well as the libraries who generously filled these requests. I would like to acknowledge IBM, which made its database on U.S. patents available on the web. I would also like to acknowledge the National Science Foundation, which has supported me through grant CCR-9731758.

I gratefully acknowledge the permission of the following publishers to reproduce copyrighted material in this book: Figures 12.1 and 17.11 are reproduced from *Australian Mathematics Teacher* with the kind permission of the Australian Association of Mathematics Teachers. Figure 21.11 is reproduced from *Recreational Problems in Geometric Dissections, and How to Solve Them,* by Harry Lindgren (1972), with the kind permission of Dover Publications, Inc. Figures 8.2 and 18.22 are reproduced from the *Journal of Recreational Mathematics,* ©1979, 1985 with the kind permission of Baywood Publishing Company, Inc. Figure 15.14 is reproduced from the *Mathematical Gazette* with the kind permission of the Mathematical Association. Figure C2 is reproduced from *Old and New Unsolved Problems in Plane Geometry,* by Victor Klee and Stan Wagon (1991), with the kind permission of the Mathematical Association of America.

I would like to thank my editor, Lauren Cowles, who has continued to supply exceptionally useful advice. She has anticipated the fascination that readers would have toward this unique and unusual topic and has steadfastly championed my books.

CHAPTER 1

if it Ain't
got that Swing

A geometric dissection is a cutting of a geometric figure into pieces that can be rearranged to form another figure. As visual demonstrations of relationships such as the Pythagorean theorem, dissections have had a surprisingly rich history, reaching back to Arabian mathematicians a millennium ago and Greek mathematicians more than two millennia ago. As mathematical puzzles, they enjoyed great popularity a century ago in newspaper and magazine columns written by the American Sam Loyd and the Englishman Henry Ernest Dudeney. Loyd and Dudeney set as a goal the minimization of the number of pieces. Their puzzles charmed and challenged readers, especially when Dudeney (1907) introduced an intriguing variation in his book, *The Canterbury Puzzles*. After presenting the remarkable 4-piece solution for the dissection of an equilateral triangle to a square, Dudeney wrote:

> I add an illustration showing the puzzle in a rather curious practical form, as it was made in polished mahogany with brass hinges for use by certain audiences. It will be seen that the four pieces form a sort of chain, and that when they are closed up in one direction they form the triangle, and when closed in the other direction they form the square.

This hinged model, in Figure 1.1, has captivated readers ever since. It is just too nifty not to have been described in at least a dozen other books in the intervening years. In (1972), the geometer and math historian Howard Eves described a set of four connected tables that swing around to form either a square or a triangular top, thus accommodating card games with either three or four players! There is something irresistible about the idea of swinging hinged pieces one way to form one figure and another way to form another figure. You do not really need a physical model to enjoy this property. Once you have examined Figure 1.1, you will be swinging mental images of the pieces around in your mind.

This dissection is *swing-hingeable* because we can link its pieces together with *swing hinges,* so that when we swing them one way on these hinges the pieces form one figure, and when we swing them around in some other way the pieces form the other figure. If we cannot use swing hinges in this way, I will call the dissection

1

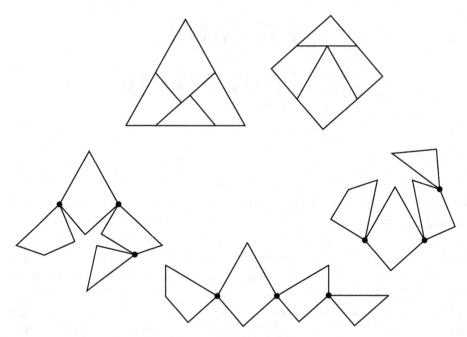

1.1: Hinged dissection of a triangle to a square

unhingeable. We call a set of pieces plus the hinges that link them together a *hinged assemblage.* For swing-hinged dissections of two-dimensional figures, which fill most of this book, we will assume that the pieces stay in the plane as we swing them around on their hinges. Thus we turn over no pieces when we use swing hinges.

Let's consider the 4-piece dissection of two equal squares to a larger square, described in Plato's *Menon* and also in his *Timaeus.* As we can see in Figure 1.2, it too is hingeable. Here there are two hinged assemblages, each of which forms one of the two equal squares. The two assemblages together form the large square. In general, a dissection of $n > 1$ figures to a single figure is hingeable if each of the n figures has its own hinged assemblage that forms the figure and if the set of n hinged assemblages together form the large figure.

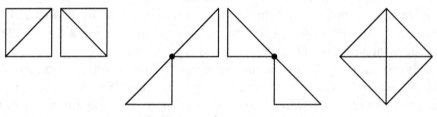

1.2: Two equal squares to one

Another example is the dissection by Sam Loyd (*Tit-Bits,* 1897b) of two Greek Crosses to a square in Figure 1.3. Since it is a dissection of two figures to one and there are two hinged assemblages, Loyd's dissection is hingeable (Figure 1.4). A similar notion applies to dissecting $n > 1$ figures into $m > 1$ figures, as long as no proper subset of the n figures is of total area equal to that of some subset of the m figures. Then the dissection is hingeable if every piece is in one of $m + n - 1$ assemblages, which together form the n figures and which also form the m figures. Examples for $m = 2$ and $n = 2$ are shown in Figures 7.2 and 8.3. If a proper subset of the n figures were of area equal to that of some subset of the m figures, then we could treat the dissection as the union of several dissections and use fewer than $m + n - 1$ assemblages.

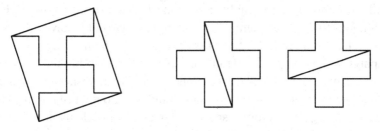

1.3: Loyd's two Greek Crosses to one square

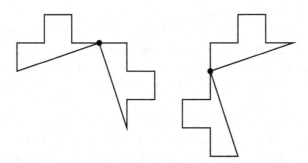

1.4: Hinges for two Greek Crosses to one square

When I wrote my first book on geometric dissections, I became fascinated with hingeable dissections. I realized that many hingeable dissections had been published without any indication that they are hingeable. I have since discovered such an abundance of hinged dissections that it is tempting to ask: Given any pair of figures that are of equal area and bounded by straight line segments, is it always possible to find a swing-hingeable dissection of them?

William Wallace (1831), Farkas Bolyai (1832), and P. Gerwien (1833) proved the analogous property for normal (unhinged) dissections: For any pair of figures that

3

are of equal area and are bounded by straight line segments, it is possible to find a dissection between them consisting of a finite number of pieces.

By way of contrast, suppose that we restrict ourselves to *translational* dissections, in which we move the pieces from one figure to another using only translation with no rotation. For example, the dissection of two Greek Crosses to a square in Figure 1.3 is translational, but the dissection of a triangle to a square in Figure 1.1 is not translational. In fact, no translational dissection between these latter two figures is possible. In (1951), the Swiss geometer Hugo Hadwiger and P. Glur gave a simple characterization of which pairs of figures have such dissections.

To see if hingeability is restrictive in the same way that translation with no rotation is, I began searching for hingeable dissections. I chose as a natural goal the minimization of the number of pieces, subject to the dissection being hingeable. As I found more and more hingeable dissections, I became preoccupied with identifying techniques to discover them. Consequently, I have not resolved the question of whether hingeability restricts the class of dissectable figures.

Compared with finding normal (unhingeable) dissections, finding hingeable dissections seems much harder – like accomplishing a difficult task with one hand tied behind your back. The standard dissection techniques do not seem to be adequate and need to be reengineered for these more challenging problems. In the end, however, suitably modified versions of techniques such as tessellations, strips, slides, steps, and polygonal structure, along with a goodly measure of determination and hard work, allow us to find a wide variety of hingeable dissections.

In time, I became so entranced with these dissections that unhingeable ones began to seem almost deficient. Appropriating the words from that famous song by Irving Mills and Duke Ellington, I started to feel that "it don't mean a thing, if it ain't got that swing" – *doo-Wah doo-Wah Doo-wah doo-Wah, doo-Wah doo-Wah Doo-wah doo-WAH!*

Was Henry Dudeney the first to get dissections to swing? Although he played a leading role in popularizing dissections as mathematical puzzles, the triangle-to-square dissection was not the first dissection recognized to be hingeable. In (1864), the British mathematician and physicist Philip Kelland wrote:

> The modification which I gave of the demonstration of [the 47th proposition of Euclid's first book] in the notes to my edition of Playfair's Geometry (edition 1846, p. 273), has had the honor of being exhibited in two different mechanical forms. The first by rotations without sliding, whereby the two squares on the sides, when placed together, are converted into the square on the hypothenuse; the second, by two transpositions (slidings) without rotation, whereby the same change is effected. The former is obvious enough, and could have escaped nobody.

4

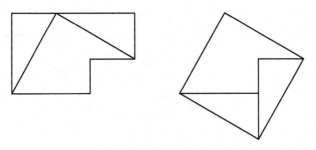

1.5: Two attached squares to one

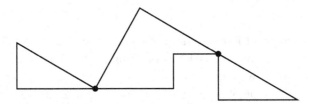

1.6: Hinged pieces for two attached squares to one

This dissection is in Figure 1.5, and one of its three possible hingings is in Figure 1.6. How remarkable that Kelland singled out the two properties: translation, as studied by Hadwiger and Glur, and hingeability, explored here! The dissection derives from a dissection of two unequal squares to one (Figure 4.4), which was discovered by the Islamic mathematician and scholar Thābit in the ninth century. The 3-piece version shown in Figure 1.5 might have come from a figure by the German academic Johann Sturm (1700), in which pieces have three different shades. In (1921), Johannes E. Böttcher, a professor and rector at the Realgymnasium in Leipzig, gave a diagram that illustrated another hinging for this dissection. Unfortunately, he drew one of the two triangles in the wrong orientation.

Other people either identified dissections as hingeable or came close to doing so. In (1905), the blind English geometer Henry Taylor described a number of hingeable dissections. His description of how to rearrange the pieces from one figure to form the other came tantalizingly close to stating that they were hingeable: He identified rotations about points as he described how to move some of the pieces from one figure to another. Figure 1.8 illustrates his hingeable dissection of an irregular triangle to another irregular triangle of equal base and height. The hinged pieces are in Figure 1.7.

In (1940), the American mathematician Robert Yates described several dissections for which he specified rotation. In (1960), Harry Lindgren, the master of dissections and an Australian patent examiner, identified the Q-slide dissection technique as leading to a hingeable dissection. In (1963), Donald Bruyr, a professor of

5

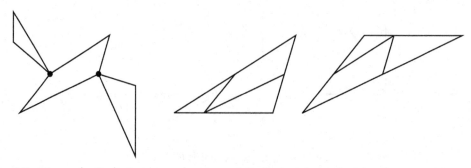

1.7: Hinges for Taylor's   **1.8:** Taylor's irregular triangles

mathematics education at Kansas State Teachers College, gave various hinged examples of dissections, which he called generating boards. In (2000b), Jin Akiyama and Gisaku Nakamura, two combinatorial geometers at Tokai University in Tokyo, initiated a study of a type of hingeable dissections that are closely related to the triangle-to-square dissection in Figure 1.1. In addition, there are a number of people who have identified hinged dissections of three-dimensional objects, which I will discuss in Chapter 20.

When I wrote my chapter on three-dimensional hinged dissections, I considered several types of hinges. Shifting back to two dimensions, I was particularly intrigued by *twist hinges,* which use a point of rotation on the interior of an edge shared by two pieces. With twist hinges, we can flip one piece over relative to the other, using rotation by 180° through the third dimension. Subsequently, I learned of an earlier dissection of this sort by William Esser III, who was awarded a U.S. patent in (1985) for what was essentially the dissection of an ellipsoid to a heart-shaped object.

Esser sliced the ellipsoid along a diagonal through its center. A rotation of 180° then produces the heart shape. The two-dimensional analogue of this dissection is shown in Figure 1.9. A small open circle indicates the position of a twist hinge. I mark each piece that is flipped an odd number of times with an "*" in the ellipse

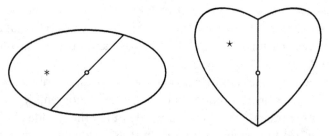

1.9: Twist-hingeable dissection of an ellipse to a heart

and with a "$\star$" in the heart. After flipping the piece, I have rotated the assemblage by 45° to highlight the symmetry of the heart.

Interestingly, Ernst Lurker, a sculptor and artist, discovered a very similar dissection around the same time. Erno Rubik (1983), the creator of Rubik's cube, also described a twist-hinged dissection, but these are the only previously published twist-hinged dissections of which I am aware.

The movement made possible by twist hinges seems rather limited. Thus I was shocked to learn that I could find a lot of twist-hingeable dissections. In addition to discovering many individual dissections, I have discovered two techniques that convert large classes of swing-hinged dissections to twist-hinged ones. It makes sense to delay the twist-hinged dissections until after we have thoroughly mastered swing-hinged dissections. However, as we swing along through the book, we will identify a crucial property in the swing-hinged dissections that we will take advantage of when we come to perform the conversions.

We will see so many twist-hinged dissections that it is tempting to pose a question similar to the one I have already posed: Given any pair of figures that are of equal area and bounded by straight line segments, is it always possible to find a twist-hingeable dissection of them? And I admit once again that I have no idea whether the answer is yes or no.

Compared with the history of normal geometric dissections, the history of hingeable dissections is rather sparse. Yet these puzzles and their corresponding demonstration pieces are so fascinating that I have sought related or comparable objects that people have studied, such as: coloring hinged models, bracing regular polygons, hinged tessellations, hinge convertibility of polyominoes, and piano-hinged polyhedral surfaces. I have distributed short discussions of these topics throughout the book, under the heading "Turnabout."

I have also crafted a bit of mystery from my doubts about whether Henry Dudeney discovered the 4-piece triangle-to-square dissection. The resulting installments, entitled "The Curious Case of the Brass Hinges," subject Dudeney's columns to statistical, logical, and even Freudian analysis! These vignettes also furnish us with clues as to the nature of the mathematical puzzle experience a century ago.

To keep readers off their mental couches, I have provided puzzles here and there, with solutions in the final chapter. Some readers may find these too difficult, in which case they can enjoy the less demanding task of verifying that the sets of hinged pieces presented throughout the book can indeed swing around to form the figures as claimed. Other readers may find the puzzles too easy, in which case I challenge them to survey the dissections in the book to find those that they can improve.

In any event, I hope that you are ready to experience all of the twists and turns of this wonderful subject. So now, baby, let's swing!

The Curious Case of the Brass Hinges

Henry Dudeney, the man who introduced the legendary hinged model of a triangle to a square, was a "pivotal" figure in the mathematical puzzle world. His first major success was his puzzle column in the *Weekly Dispatch*, "Puzzles and Prizes," which began on April 19, 1896 and ran through the end of 1903. For most of this period he published his column weekly and awarded monetary prizes for the best solutions. Dudeney allowed a week for contestants to send in their solutions, and he announced his solution and made the awards two weeks after the puzzle had originally appeared. Many of the puzzles in his books, *The Canterbury Puzzles* and *Amusements in Mathematics,* first appeared in his column in the *Dispatch.* He was the first to pose the puzzle of dissecting an equilateral triangle to a square, on April 6, 1902:

> It may have struck the reader as a little curious that amidst all the dissection problems that have appeared in these columns and elsewhere, the most obvious one of all - the conversion of the equilateral triangle into a square - has never (so far as I know) been dealt with. I say most obvious, because the triangle and the square are the simplest of all symmetrical rectilineal plane figures. I have certainly had the problem in my notebooks for some years, but have kept it back principally on account of its inherent difficulty. As our readers have now had some experience in such matters we will consider the puzzle and see what they can make of it. I am confident that it will be solved. But it will be interesting to see how many will get at the heart of the mystery in the fewest possible pieces.
>
> No. 440. – The Triangle and the Square.
>
> The puzzle is to cut any equilateral triangle (that is, a triangle whose three sides are of equal length) into as few pieces as possible that will fit together and form a perfect square.

Curiously, however, Dudeney did not give the solution two weeks later (on April 20) as expected. Instead, he wrote:

> Several competitors correctly judged that if five were the smallest possible number of pieces into which an equilateral triangle may be cut to form a square I should not have written of the "inherent difficulty" of the problem. To cut it in five pieces all that you have to do is to first divide your triangle into two equal right-angled triangles by a cut perpendicular to one of the sides and then place these two pieces together so as to form a rectangle. Now cut this rectangle so as to form a square. If the rectangle were whole this could be done in three pieces, but the diagonal cut already made necessitates our producing

five pieces. This method involves turning over (which was not prohibited), but it gives the key to a solution in five pieces without turning over any piece.

The correct solution, however, is much prettier, and is in only four pieces, and, surprising as it may seem, it is not necessary to turn over any piece. It seems that I was right in believing that it would be solved, though only one competitor succeeded in getting at "the heart of the mystery." This was Mr. C. W. McElroy, 25, Great Jackson-street, Hulme, Manchester, to whom the prize of half a guinea will be sent.

As this is, perhaps, one of the most interesting, if not important, of the over four hundred problems that have appeared in these columns (the prize winner calls it "a little beauty") I have decided to withhold the solution for a fortnight, in order that readers may have another look at it. Any correct solutions received up to the 28th inst. will be duly acknowledged in our issue of May 4 next. The difficulty consists merely in conceiving the initial principle; that done, all is comparatively easy. So far as I know there is only the one solution.

I should like on this occasion to say that Mr. McElroy, whose name has appeared so frequently in these columns, is, I consider, one of the ablest all around puzzle-solvers in this country. His mastery of the above little problem would not, of course, alone entitle him to any such appreciation, but for over five years there can hardly have been more than a few weeks in which he has not sent me his answers to the puzzles, and one is therefore fully competent to express an opinion on his work. I think his first communication was in the early part of 1897, when he sent me a correct solution to the "Oriental Problem" in *Tit-Bits,* one of the most formidable puzzles ever inflicted on a long-suffering public, wherein a merchant of Bagdad dealt out honey to a messenger from the Caliph, and water to his camels, a process entailing over 500 most intricate manipulations. Yet I have not the slightest idea as to the age of Mr. McElroy, nor as to his vocation in life. He is simply known to me as a puzzle-solver, and if I have occasion to consider whether a particularly hard nut is likely to be cracked, I always bear in mind that clever worker down at Manchester.

Finally, Dudeney gave the solution two weeks after that, on May 4. He first described a method with ruler and compass to locate the dissection cuts. Then he stated:

> The key to the solution lies in the fact that there are three angles of 60deg. to be got rid of and four right angles to be made. The three angles of 60deg. will, if added together, make 180deg., so if the cuts are so made that these three angles meet at a point they will form a straight line. The diagram shows how this is done.

The extension of time has resulted in not a single correct solution being received, so the puzzle may be regarded as a decidedly hard nut.

At first glance, it would seem that Dudeney already knew the 4-piece solution before posing the puzzle. Yet a careful reading of the columns and a consideration of the extraordinary circumstance of an extension conjures up another possibility: Dudeney may not have known of the 4-piece solution until he opened McElroy's letter. He may only have let it appear that he was already familiar with it.

Perhaps Dudeney did not supply the solution on April 20 because he did not yet understand McElroy's solution well enough to explain it. In contrast, he was prepared to explain a 5-piece solution. Also, Dudeney posed the problem in the open-ended form of finding the fewest possible number of pieces. If he knew it was so challenging, why did he not immediately let the readers know it could be done in four pieces? (He did do this when he posed the problem in *The Canterbury Puzzles*.) Dudeney did not always have a solution to problems posed, as demonstrated by this quote from "Our 500th Puzzle," in "Puzzles and Prizes," June 7, 1903:

> As often happens, I had given a problem that I had not at the time myself solved ....

So here is our curious case: Did Henry Dudeney discover the 4-piece dissection of a triangle to a square, which he so brilliantly hinged later on? Or did he simply appropriate it from McElroy?

CHAPTER 2

Hinging on the Details

Let's take a moment to consider details related to swing-hinged dissections of two-dimensional figures, which constitute most of this book. We start by constructing physical models and then go on to explore variations of swing hinges, patterns of hinging pieces, and properties of swing-hinged dissections. We will consider in what sequence to swing the pieces and conclude with a description of a mathematical model.

Physical models of swing hinges. Of course, any physical model must have three dimensions. We shall assume that the third dimension in our model is relatively thin. The easiest way to produce a physical model of a hinged dissection is to cut the pieces out of sturdy paper or cardboard. For hinges, take short pieces of thread and tape them to the underside of the pieces. Figure 2.1 shows an example, using the triangle-to-square dissection of Figure 1.1. Wherever we need a hinge between two pieces, we use a piece of thread, held in place by two pieces of tape (shown shaded). To discourage turning pieces over, mark the underside of each piece with an X. This is a good method for classroom use.

If you desire a sturdier construction, cut the pieces out of sheets of Plexiglas or thin wood. Hinge the pieces using a tape whose width is the thickness of the pieces, and place the tape on the thin edges of the pieces on each side of any hinge point. For use in talks with an overhead projector, I have had a number of dissections

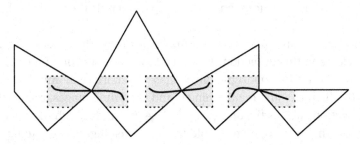

2.1: Hinged model of a triangle to a square using paper, thread, and tape

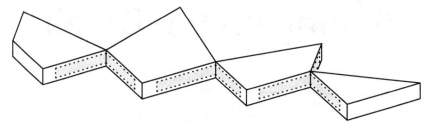

2.2: Hinged model of a triangle to a square using Plexiglas and tape

precisely cut out of Plexiglas using a laser cutter and have hinged them using a thin black tape. Because the pieces are thin and the tape is flexible, we need to handle such hinged assemblages with care. Figure 2.2 shows a model of the triangle-to-square dissection in perspective.

For more rugged use, I recommend cutting the pieces out of a thicker wood and using metal hinges to connect them. This method requires good craftsmanship, since otherwise the pieces will not fold together properly. The hinge consists of three parts. One part is attached (generally screwed) to one of the dissection pieces, a second part is attached to the other dissection piece, and the third part is a pin centered on the axis of rotation and surrounded by a sleeve that wraps around the pin and is shared by the two other parts. The first part includes the first, third, fifth,… sections of the sleeve, and the second part includes second, fourth, sixth,… sections of the sleeve.

Variations of swing hinges. Sometimes, as in Figure 10.22, it is useful to have three pieces share a hinge, which I term a *triple hinge.* When three pieces share a hinge, the relative rotational position of the three pieces is important. We shall assume that these positions do not change: Piece 1 comes before piece 2, which comes before piece 3, in clockwise order around the hinge.

Triple hinges have simple physical realizations. In the tape model, three pieces of tape suffice. In the hardware model, a triple hinge would have the first part include the first, fourth, seventh,… sections of the sleeve, the second part include the second, fifth, eighth,… sections, and the third part include the third, sixth, ninth,… sections.

In principle, the only limit to the number of pieces that share a hinge is the number of pieces in the dissection. I found a hinged dissection of a Latin Cross to a dodecagon that used a quadruple hinge; alas, the more economical dissection in Figure 15.26, which has just regular hinges and two triple hinges, replaced it.

In a dissection such as the one in Figure 12.4, two different hinges abut against each other. In other words, when we fold together the hinged pieces to make one of the figures, the axes of rotation of the two hinges coincide. In the tape model, the tape is thin enough so that no problem arises. In a hardware model of such hinges, we can have the two sets of hinges share the vertical column. The one set would use two pins in the first and third quarters of the column, and the other set would use two pins in the second and fourth quarters of the column.

Hinge patterns. Let's characterize patterns for hinging pieces. If the pieces form a simple chain, then the pieces are *linearly hinged.* The dissection of a triangle to a square in Figure 1.1 is linearly hinged, as is the dissection of two attached squares to one (Figure 1.6).

A hinged assemblage is *cyclicly hinged* if the pieces and hinges form a cycle and if each hinge is shared by just two pieces. Whenever a cyclicly hinged assemblage has a hole into which we wish to move another assemblage, we must necessarily use the third dimension when we move pieces from one figure to another. Chapter 9 contains many such examples.

David Eppstein, a computer science professor at the University of California at Irvine, recognized that if there is a hingeable dissection of one figure to another then there also is a cyclicly hingeable dissection of those figures. Here is how you can convert a hingeable dissection into a cyclicly hingeable dissection, assuming that there are no cycles in the given dissection: For each piece that has only one hinge on it, make a cut from the hinge point to an opposite side and place a hinge at that point. For each piece that has two or more hinges on it, choose a point in the interior of the piece and make a cut (not necessarily a straight line segment) from each hinge to that interior point. Split each original hinge into k hinges, where k is the number of cuts from that hinge. Necessarily, the k hinges will abut against each other.

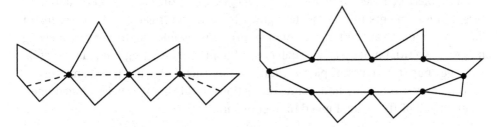

2.3: Converting the hingeable triangle-square to be cyclicly hingeable

The conversion of the hinged assemblage in Figure 1.1 is in Figure 2.3. On the left is the original hinged assemblage, augmented with dashed edges to show the new cuts. For each piece that had two hinges on it, I chose a point in the interior midway between the two hinges. On the right is the resulting cyclicly hinged assemblage. Of course, any cyclicly hingeable dissection is also linearly hingeable.

(Just split the cycle at some hinge!) Thus, if a dissection of one figure to another is hingeable, then there is a linearly hingeable dissection of those figures.

Properties of hinged dissections. Suppose that we wish to cut a dissection out of a piece of wood that has a grain running in one direction. Certainly we can cut the pieces so that the grain runs in the same direction when we swing the pieces to form one of the figures. If the grain is still lined up after we swing the pieces to form the other figure, then we say that the dissection is *grain-preserving.*

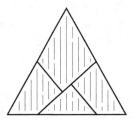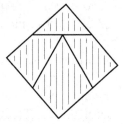

2.4: Dissection of a triangle to a square is grain-preserving

The swing-hinged dissection of a triangle to a square in Figure 1.1 is grain-preserving, as we see in Figure 2.4. On the other hand, the dissection in Figure 1.6 is not grain-preserving. However, there does exist a 4-piece hinged dissection of two attached squares to one that is grain-preserving. Can you find it? (*Hint:* It is related to the hinged dissection in Figures 4.1 and 4.3.) Twist-hinged dissections can be grain-preserving, too. When a twist-hinged dissection is grain-preserving, the grain must run in one of just two directions. Can you find them in Figure 1.9?

Two pieces connected by a hinge are *hinge-snug* if they are adjacent along different line segments in each of the figures formed and if each such line segment has one endpoint at the hinge. This key property enables us to convert a swing hinge to two twist hinges, as we shall see in Chapter 22. A swing-hinged dissection is hinge-snug if all pairs of pieces connected by each hinge are hinge-snug. Thus we can convert the many hinge-snug dissections in this book to twist-hinged ones, including all of the swing-hinged dissections in Chapter 1.

Motion planning. Given a hinged dissection, how do we determine if there is an order in which to swing the pieces from one figure to the other, so that the pieces stay in the plane and no piece crosses over another? It does not seem to be easy to determine if such an order exists. Indeed, it may be necessary to move several pieces simultaneously in order to perform the transformation.

A basic question is whether one piece can be cleanly separated from another piece in the same assemblage by rotation about a specific hinge. Let's assume that

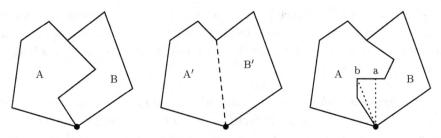

2.5: Determining if two pieces can be separated: simple cases

the boundary of each piece consists of straight line segments. Let's also assume that the two pieces (A and B) are not convex, because if they are convex then we can always separate them.

If replacing their common border by a straight line segment yields two convex pieces A′ and B′, then A and B are separable if and only if the distance from the hinge is nondecreasing as we move along the common border away from the hinge. On the right and the left of Figure 2.5 are pairs of pieces A and B, attached by a hinge. As shown in the middle, replacing the common border in each case gives a straight line. Pair A and B on the left are separable. However, pair A and B on the right are not separable because the distance from the hinge has decreased as we have moved from point b to point a. If replacing the common border of A and B by a straight line segment does not yield two convex pieces, then things are trickier but can still be checked. Let's skip those details.

Some hinged dissections have the drawback that we cannot swing the pieces in an assemblage from one figure to the other while staying within two dimensions. We need to lift one piece so that it slides over another and then drops into place. (Figure 7.2 is an extreme example of this.) Whether we should allow this is a matter of taste. Should we require the pieces of any assemblage to stay in two dimensions as we rotate them? In a physical model, if the pieces are thin enough and the hinges are ever so "wobbly," then we can perform the lifting up and sliding over whenever all points of obstruction are at least at some minimum nonzero distance from the hinge. We call such dissections *wobbly hinged*. Often, we can convert a wobbly hinged assemblage to be a nonwobbly one by cutting one of the pieces into two pieces and connecting them with a hinge.

Other types of hinges. Besides swing hinges and twist hinges, other types of hinges are possible. I do discuss two more possibilities in Chapter 21. However, I have focused specifically on swing hinges and twist hinges, because they seem more fundamental.

In three dimensions, the analogue of a swing hinge is a *piano hinge,* like the hinge that attaches the lid to the piano. The analogue of a twist hinge, which we have seen in Figure 1.9, is a *socket hinge.* For that type of hinge, one piece has a pin protruding from it, which fits into a sleeve in the other piece. The first piece rotates (about the axis of its pin) relative to the second piece; an example of this hinge is shown in Figure 20.8.

Some simple notation. Here is some simple notation for two-dimensional figures. A *polygon* is a closed plane figure bounded by straight line segments. The simplest type of polygon is a *regular polygon,* in which all sides are of equal length and all angles are equal. The notation $\{p\}$ represents a regular polygon with p sides, for example:

triangle: $\{3\}$	hexagon: $\{6\}$	enneagon: $\{9\}$
square: $\{4\}$	heptagon: $\{7\}$	decagon: $\{10\}$
pentagon: $\{5\}$	octagon: $\{8\}$	dodecagon: $\{12\}$

We represent a p-sided polygon that is not regular by $\{\tilde{p}\}$. For example, $\{\tilde{4}\}$ is an unrestricted 4-sided polygon or *quadrilateral.*

Other favorite shapes include *star* polygons. We form the boundary of a regular polygon $\{p\}$ by placing p vertices at equal intervals around the circumference of a circle and connecting each vertex to its clockwise neighbor with a straight line segment. If $p > 4$ then we can generate a star, denoted by $\{p/2\}$, by connecting each vertex not to its neighbor but to the second nearest vertex in a clockwise direction. We take the star to be the area enclosed within the line segments and take the boundary to be those portions of the line segments that separate the star from the rest of the plane. The $\{5/2\}$, also called a *pentagram,* appears in Figure 9.15. The $\{6/2\}$, also called a *hexagram,* appears in Figure 3.25. In general, for $1 < q < p/2$ we have a star $\{p/q\}$, where we have connected each vertex to its qth nearest vertex in a clockwise direction. Examples of $\{8/3\}$s are in Figure 9.19.

As for other shapes, we have already seen in Figure 1.3 the Greek Cross, which we designate with $\{G\}$. Another cross is the Latin Cross, or $\{L\}$, which makes an appearance in Figure 4.33.

Mathematical model. Now that we have put forward an informal notion of swing-hinged dissections, let's propose an appropriate mathematical model. Readers not familiar with the notions of open and closed sets may want to skip to the next chapter.

When we make a cut, how should we describe the resulting two edges precisely? Does one of the resulting pieces get the points on the cut line, making the other piece look like an open set along the corresponding edge?

In order to derive a reasonable mathematical model, let's retreat to physical reality: If we cut a piece of wood, then we "consume" the portion of the wood along

16

the cut line; that is, we convert it to sawdust. Even using a precision cutting device such as a laser consumes some portion of material. Although I haven't looked under the microscope to see what happens to paper when I cut it with scissors, I suspect that the cut line is also consumed.

So here is a mathematical model for cutting: Assume that a figure such as a polygon or a star is an open set. That means that its boundary is not part of the figure itself. When we cut the figure along a sequence of line segments, we effectively remove all points on those line segments, resulting in pieces that are open sets.

When we assemble two pieces, we could imagine gluing them together by adding their common boundary, minus the endpoints of the common boundary. If we use real glue, then we could not pull the pieces apart. If we do not use real glue but just move the pieces up against each other, the "glue" would be a thin "film" of air molecules. If this film were not there, then a suction would prevent us from pulling the pieces apart. In either case, the assembled figure is an open set.

We represent a hinge by the point at the center of rotation of the pieces that the hinge attaches. Since the pieces are open sets, a hinge will not obstruct the movement of the pieces. The only problem comes if we have two hinges that abut against each other. We then need to allow two different hinges to occupy the same point in the plane. Since this is not much worse than assuming that one hinge occupies no area at all, we allow this too!

In this chapter, we have explored physical and mathematical models of hinged dissections and have studied properties of hingings. We are almost ready to swing into action, but first need some tricks to help us design our hinged dissections. We turn our attention to these in the next chapter.

CHAPTER 3

Tricks for Turning

So far, we have relied on the kindness of old friends, identifying previously known dissections that we could hinge. Yet sooner or later we must produce our own hinged dissections in a world more disposed to turning a cold shoulder than to swinging to our assistance. How are we to design new hingeable dissections or even explain the properties that make the old ones hingeable? Serendipitously, we can fall back on techniques such as tessellations, strips, slides, steps, and polygonal structure. Let's take stock of these techniques and see how to turn them to our advantage.

We start with tessellations. Given a plane figure, a *tessellation of the plane* is a covering of the plane with copies of the figure in a repeating pattern such that the copies do not overlap. The figure that we use to tile the plane is a *tessellation element* and consists of one or more pieces. For example, a square (or, more generally, any quadrilateral) tiles the plane. So does a triangle, and also a regular hexagon. There are many different tessellations of the plane.

The technique of *superposing tessellations* is a powerful tool in dissecting figures. We take two tessellations with the same pattern of repetition and overlay them in a way that preserves this common pattern of repetition. The line segments in one tessellation induce cuts in the figures of the other and vice versa. We can derive many dissections by superposing tessellations. Of course, there is no guarantee that the resulting dissections are hingeable.

With one more restriction we can produce dissections that are hingeable. A tessellation has *rotational symmetry* about a point if we can rotate the tessellation by some angle smaller than 360° about that point so that the resulting figure coincides in every detail with the original. The tessellation possesses *n-fold rotational symmetry* if the angle of rotation is $360°/n$. Call a point about which there is rotational symmetry a *symmetry point.* Suppose we superpose two tessellations so that they share no line segment of positive length. Suppose also that a line segment in one tessellation crosses a line segment in the other only at a symmetry point in each tessellation. Then there is a dissection derivable from the superposition that will be hingeable, because we can introduce hinges at those points in the dissection.

18

An example is a dissection of a Greek Cross to an isosceles right triangle. Henry Dudeney (*Tit-Bits,* 1897a) gave a pretty 4-piece hingeable dissection (Figure 3.2), though there is no evidence that Dudeney knew the dissection to be hingeable.

Using the tessellation elements in Figure 3.1, we derive the dissection by overlaying the tessellations in Figure 3.3. Small dots indicate the symmetry points. Those at the midpoints of the hypotenuses of the isosceles right triangles have 2-fold symmetry, whereas the others have 4-fold symmetry. Placing hinges at the symmetry points, we get the hinged pieces in Figure 3.4.

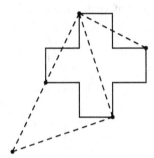

3.1: Tessellation elements

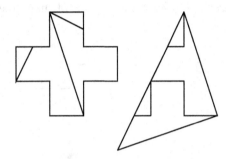

3.2: Cross to an isosceles right triangle

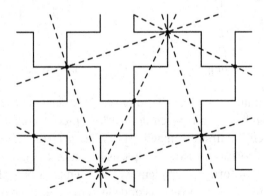

3.3: Superposed tessellations of a Greek Cross and an isosceles right triangle

The sharp-eyed reader will see that there is a pair of line segments in Figure 3.3 that cross at a point that is not a symmetry point in either tessellation. Can you find it? Luckily there are enough symmetry points to provide for three hinges, which is all that we need in this case. Thus, our requirement – that each crossing of line segments be at a common symmetry point – is sufficient but not necessary to produce a hingeable dissection. In Chapter 11, we will identify more such exceptional examples.

19

Curiously, there is a second way to hinge the pieces (Figure 3.5), which turns out to be hinge-snug. We hinge each of the two triangles at the vertex at which the short side and the hypotenuse meet. These two hinges are not located at symmetry points of the tessellations. So why does this work? The answer is that these points are a *linked pair of local symmetry points* in the superposed tessellations. They each have rotational symmetry for a neighborhood around the points in the following sense. The points are one fourth and three fourths of the way along the hypotenuse of the isosceles right triangle. Each point is in the center of a zigzag line and has a dashed line through it. The pieces on each side of these lines are different, so that we need a pair of them to effect a swap of small triangles.

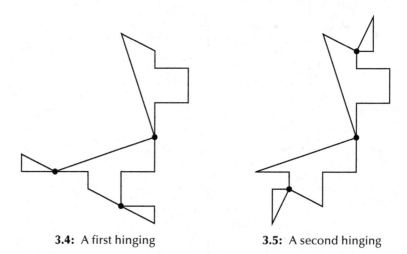

3.4: A first hinging   3.5: A second hinging

When the possibilities for tessellations seem exhausted, where will we turn? Let's try our old friend, the T-strip technique. The 4-piece triangle-to-square dissection in Figure 1.1 seems at first like magic. Harry Lindgren (1951) showed how to derive it from the crossposition of two T-strips, as in Figure 3.6. In this method, we cut a figure into pieces that form a strip element. We then fit copies of this element together to form a strip but rotate every second element in the strip by 180° and match it with an unrotated twin. Thus we can call it a *twinned-strip,* or *T-strip.* Since every second element is rotated, every two consecutive elements in the strip share a point of 2-fold rotational symmetry. Let's call such points *anchor points.* We can similarly create a T-strip for the other figure. We then crosspose the two T-strips, making sure that an anchor point in one strip that is covered by the other strip either overlays an anchor point in the other strip or falls on a boundary edge of the other strip.

In the crossposition of the triangle and the square in Figure 3.6, the two T-strips overlap on an area that is double the area of either strip element. Lindgren (1964b) called this a *TT2* dissection. When the area of overlap equals the area of the strip

20

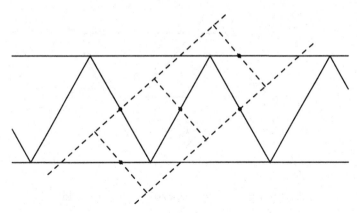

3.6: Crossposition of triangles and squares

element, Lindgren called it a *TT1* dissection. If we can form a strip without rotating every second element and furthermore do not use anchor points in the strip, then we call the strip a *plain-strip,* or *P-strip.* It is possible to crosspose a P-strip with either a P-strip or a T-strip, giving *PP* and *PT* dissections, respectively.

The T-strip technique gives us a tool for creating hingeable dissections. Let us take as an example a dissection of one trapezoid to another trapezoid that has different angles and different heights. For the case when the trapezoids are not too dissimilar, the Cambridge mathematician William Macaulay (1919) gave a 4-piece dissection that is both hingeable and derivable by T-strips, although there is no indication that he realized that his trapezoid dissection had either property. The hinged pieces are in Figure 3.7 and the dissection is in Figure 3.8. We generate it with the TT2 crossposition in Figure 3.9. Small dots indicate the anchor points. The dissection is hinge-snug and grain-preserving.

The T-strip in the dashed edges is what we would first imagine for a T-strip consisting of trapezoids, lining up their parallel edges with the strip boundaries. By contrast, the T-strip in the solid lines has each trapezoid touch just one boundary edge of the strip. Even so, we can proceed in a linear order through the strip, with every second trapezoid rotated 180° degrees relative to the others. By inspection,

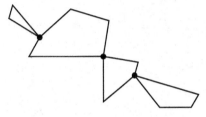

3.7: Hinges for trapezoids

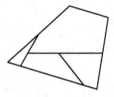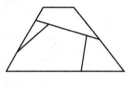

3.8: Trapezoid and a trapezoid

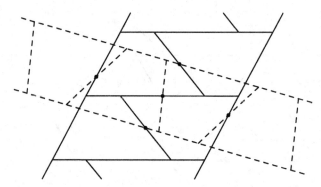

3.9: Crossposition of trapezoids and trapezoids

we can hinge the pieces at the vertices that correspond to the anchor points. Why is this possible? Furthermore, other points for hinges are at the intersection of the edge boundaries for the two strips, or where an anchor point of one strip falls on the boundary of the other. Why?

Macaulay (1922) observed that the strip technique is a type of tessellation method. The crossposition of the two strips induces two tessellations and their corresponding superposition. You can line up multiple copies of the same strip to fill out the plane. If you shift the strips relative to each other by an appropriate offset and do this for both of the strips that you are crossposing, then the resulting superposed tessellations produce the desired dissection. Figure 3.10 illustrates

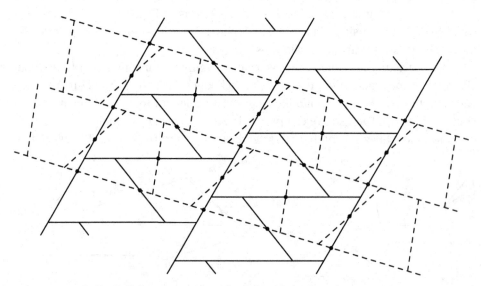

3.10: Building a superposition of trapezoids and trapezoids

this property, with two copies of each strip offset by an appropriate amount. Now we see that the anchor points of the T-strips become symmetry points of the tessellations. Likewise, we get a shared symmetry point from the intersection of a boundary of one strip with an anchor point or a boundary of the other strip. Thus we justify our use of hinges at the anchor points and boundary intersections. We will represent symmetry points by dots in the central part of the figure.

Jin Akiyama and Gisaku Nakamura (2000b) also recognized the importance of crossing tessellations at the midpoints of certain line segments in order to find hingeable dissections. But without the characterization provided here, their method can fail to produce a hingeable dissection when we attempt a TT2 dissection of a very long and thin trapezoid and a reasonably fat trapezoid. Also, these authors (2000a) focused on properties specific to the 4-piece dissection of a triangle to a square, such as having the pieces be linearly hinged and having interior and exterior edges change roles when one figure swings around to form the other.

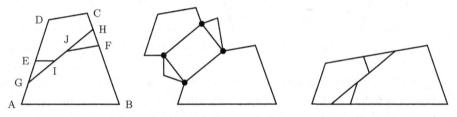

3.11: Lindgren's Q-swing dissection of quadrilaterals

When tessellations and T-strips cannot help us, we may find ourselves in a downward slide. Let's make the best of it and explore ways to make our slide elegant. In the end, we will find a way to step back up, too. Lindgren (1956) introduced the *quadrilateral slide*, or *Q-slide*, and he later (1960) gave a complete discussion. As in Figure 3.11, the Q-slide transforms one quadrilateral to another with the same angles. Lindgren positioned point E on side AD so that line segment DE is the desired length of the edge on the left. He also positioned point F on side BC so that line segment BF is the desired length of the edge on the right. Finally, he located the midpoint G of line segment AE and the midpoint H of CF. He then made a cut from G to H, a second cut from E to line segment GH that is parallel to AB, and a third cut from F to line segment GH that is parallel to CD. As Lindgren observed, the pieces are cyclicly hinged. To emphasize that the dissection is swing-hingeable, we rename the technique the *Q-swing*. This dissection is also hinge-snug and grain-preserving.

Unbeknownst to Lindgren, the basic technique for the Q-slide had appeared much earlier, in the anonymous Persian manuscript *Interlocks of Similar or Complementary Figures,* which Alpay Özdural (2000) has analyzed and dated to around

1300. This manuscript describes the conversion of a rectangle to a square in a manner consistent with the Q-slide. On the basis of references in this manuscript and similarities to a manuscript by Abū Bakr al-Khalīl al-Tājir al-Rasadī, Özdural has conjectured that Abū Bakr al-Khalīl was the originator of this technique.

When the line segment from E extends all the way to point H, then the line segment from F extends all the way to point G. In this special case, the two triangles lie flush against each other along the full extent of GH. We thus do not need to cut them apart, as Lindgren (1960) noted. A dotted line indicates the unneeded cut in the 3-piece hingeable dissection in Figure 3.12.

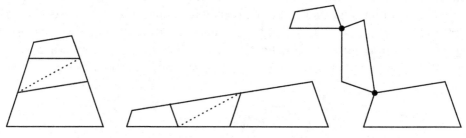

3.12: Lindgren's special case of a Q-swing

The 3-piece dissection in Figure 3.12 is less of a special case than it first appears. Whenever the length of the slide cut GH is an odd integral multiple of the length of a triangle's side along GH, there is a 3-piece dissection. Figure 3.13 illustrates the case in which GH is three times the corresponding side of the triangle. Starting above the leftmost triangle, place congruent triangles (dotted lines) along GH, alternately below and above it. Merge each of these equal triangles with the piece adjacent to and on the other side of GH. This yields the 3-piece *Q-step* dissection in Figure 3.14. We save one piece but lose one point at which to hinge. However, since the Q-swing had a spare hinge point anyway, the Q-step dissection is hingeable.

This dissection is grain-preserving. I believe that Harry Lindgren would have been pleased to see his special case so extended. Incidentally, why must the length of GH be an odd multiple? If the length were an even multiple, then the rightmost dotted triangle would either partially overlap the triangle on the right or would be

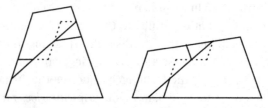

3.13: Derivation of a Q-step from a Q-swing

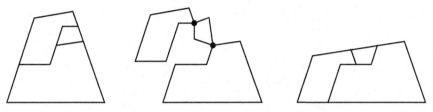

3.14: Q-step dissection applied to quadrilaterals

coincident with the triangle on the right. In either case, we would not achieve a 3-piece hinged dissection.

Starting with the Q-swing in Figure 3.13, there is another way to get a 3-piece Q-step dissection.

Puzzle 3.1 Find a second 3-piece hingeable dissection of the quadrilaterals in Figure 3.13.

In (1989), Anton Hanegraaf, a structural engineer in the Netherlands and a leader in the Dutch Cubists Club, introduced a technique that I have called the *trapezoid slide*, or *T-slide.* Hanegraaf found that it could be used to transform a trapezoid into a parallelogram when the two figures have an equal angle. Actually, the technique is more general, capable of transforming one trapezoid to another whenever they have an equal angle – as long as neither has a height more than twice the height of the other. Figure 3.15 contains such an example. Since a T-slide dissection is swing-hingeable, we rename the technique the *trapezoid-(fixed angle)-swing,* or *T(a)-swing.* As with the Q-swing, the pieces are cyclicly hinged, and the dissection is hinge-snug and grain-preserving.

Just as we can convert the Q-swing to a Q-step dissection, so can we convert the T(a)-swing. The length of the slide cut GM must be an integral multiple of the

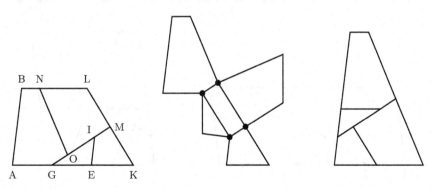

3.15: Hanegraaf's T(a)-swing dissection applied to trapezoids

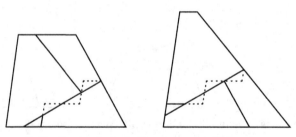

3.16: Derivation of a T-step from a T(a)-swing

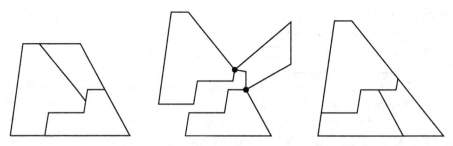

3.17: T-step dissection applied to trapezoids

length of the triangle's side along GM. Figure 3.16 illustrates the case in which GM is four times the corresponding side of the triangle. Placing congruent triangles (dotted lines) along GM and then merging these triangles with appropriate pieces gives the *T-step* dissection in Figure 3.17.

Noting all of the similarities between the T(a)-swing and the Q-swing, we might ask how the T(a)-swing relates to the Q-swing. Remarkably, we can derive the T(a)-swing from the Q-swing, as I show in Figure 3.18. Take two copies of the first trapezoid, and identify the nonparallel side that is not adjacent to the equal angle. Match the two copies of the trapezoid together along this edge, forming a parallelogram. Do the same thing with the second trapezoid, yielding a second parallelogram with

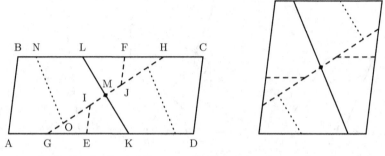

3.18: Derivation of Hanegraaf's T(a)-swing from a Q-swing

the same angles as the first. Perform a Q-swing to transform the first parallelogram to the second. This procedure cuts each trapezoid as in a T(a)-swing. In the figure, solid lines indicate the trapezoids, dashed lines indicate the Q-swing cuts, and dotted lines indicate the cuts induced by the edge along which we match the two copies.

The dissection works because the matched trapezoids have a point of 2-fold rotational symmetry, which is indicated by the dot. When applying the Q-swing to a parallelogram, the cuts of the Q-swing also have 2-fold rotational symmetry about this point. Thus, we cut each of the two copies of the first trapezoid identically; the symmetry about this point and a similar one in the second parallelogram make the additional hinges possible.

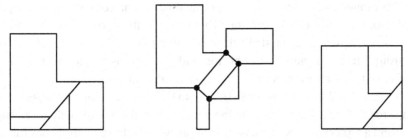

3.19: Hanegraaf's T(a)-swing dissection of a gnomon to a square

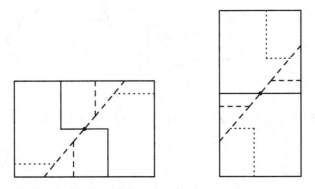

3.20: Derivation of Hanegraaf's gnomon to a square

We can match together figures more general than trapezoids, leading to a more general version of the T(a)-swing. Two copies of such a figure must match together to give a parallelogram. An example of such a figure is the gnomon produced by cutting one square out of a corner of a square whose area is four times as large. After I showed Anton Hanegraaf Figure 3.18 and the strip dissection of a gnomon to a square (Figures 11.1 and 11.2), he found the 4-piece cyclicly hinged dissection of a gnomon to a square in Figure 3.19. As we can see in the derivation in Figure 3.20,

27

two copies of the gnomon match to form a rectangle. In this case, the copies match along a (rotationally symmetric) sequence of edges instead of along a single edge, but this causes no problem as long as we cut across the sequence of edges only once. The dissection is hinge-snug and grain-preserving. Interestingly, there is more than one way to use a T(a)-swing to produce a 4-piece dissection.

Puzzle 3.2 Find another 4-piece cyclicly hinged dissection of a gnomon to a square.

The T(a)-swing also works for finding a hingeable dissection of a hexagon to a square. Anton found the 6-piece hinged dissection in Figure 3.21. Four of the pieces are cyclicly hinged, and we can hinge the remaining two pieces onto the other pieces in more than one way. This dissection is also hinge-snug and grain-preserving; the derivation is in Figure 3.22. First cut the hexagon into three hingeable pieces, which swing around to form the left and bottom sides in the rectangle on the left. The remainder of the derivation is then analogous to that of the gnomon.

In Figure 1.8 we have already seen Henry Taylor's 3-piece hingeable dissection of a triangle to another triangle with the same base and same height. William Macaulay (1914) found a 4-piece hingeable dissection (Figure 3.23) that accomplishes the same

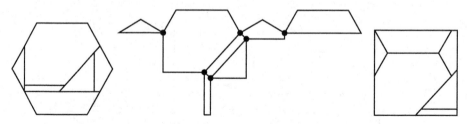

3.21: Hanegraaf's T(a)-swing dissection of a hexagon to a square

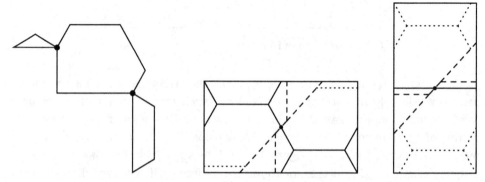

3.22: Derivation of Hanegraaf's hexagon to a square

task yet has an additional feature: piece A has its apex at the apexes of both triangles, and piece B extends along the base in each triangle. This feature allows us to maintain a hinged connection with other pieces if we alter a triangle within a larger dissection. I have used this trick in Figure 18.33. Robert Yates (1940) specified the rotation of two of the pieces in Figure 3.23. In fact, the dissection is cyclicly hingeable, hinge-snug, and grain-preserving.

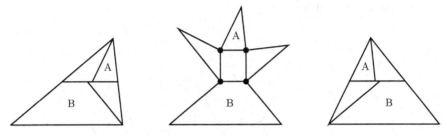

3.23: Macaulay's triangle to another of equal base and height

If we view a triangle as a trapezoid whose upper base has length 0, then Macaulay's dissection turns out to be a special case of a dissection of a trapezoid into another trapezoid in which each base maintains its length (Figure 3.24). Piece A extends along the upper base in each trapezoid, and piece B extends along the lower base in each trapezoid. This feature also allows us to maintain a hinged connection with other pieces, as we shall see in Figures 5.7 and 12.7. The line that separates piece A from piece B connects the midpoints of the two sides that are not bases. We name this technique the *trapezoid-(fixed bases)-swing,* or the *T(b)-swing.* Again, the dissection is cyclicly hingeable, hinge-snug, and grain-preserving.

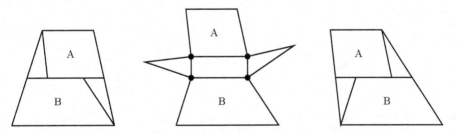

3.24: T(b)-swing: trapezoid to trapezoid with length-unchanged bases

When confronted with a challenging problem, a respectable strategy is to draw on one's inner strength. Regular polygons have an internal structure composed of rhombuses. We can exploit such internal structure in various dissections. For example, we can decompose a hexagon into three 60°-rhombuses and a hexagram into

29

six 60°-rhombuses. Thus there should be a simple dissection of a hexagram to two hexagons.

Lindgren (1964b) gave a 4-piece unhingeable dissection of a hexagram to two hexagons. We can adapt his approach to give the 6-piece hinge-snug dissection in Figure 3.25, with the two identical hinged assemblages on the left. The number 2 indicates that there are two copies of the assemblage that form the two hexagons. The letters that label the pieces indicate how the hinged rhombuses swing together to form the hexagons and the hexagram.

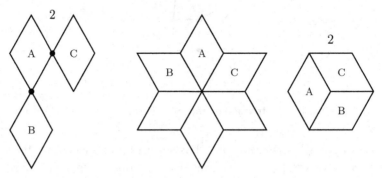

3.25: Hinged dissection of a hexagram to two hexagons

Having taken stock of our resources – namely, tessellations, strips, swings, steps, and polygonal structure – we are now able to turn challenging situations to our advantage. To celebrate, in Chapter 4 we will kick off our shoes and dance.

Turnabout 1: Coloring a Hinged Model

Suppose that you have constructed a hinged model of a hingeable dissection. Now you want to paint the pieces so that no two pieces that share a common border have the same color. This problem is related to the famous four-color problem, for which Kenneth Appel and Wolfgang Haken (1977) proved that four colors are sufficient to color a planar map. In a geometric dissection, the same set of pieces forms two different geometric figures; thus you must satisfy the constraints imposed by two maps rather than one. How many colors are necessary, and how do you find a coloring using that number of colors?

Percy Heawood (1890) studied the coloring problem for the case when each country consists of m disjoint regions. This is called the *empire*, or *m-pire*, problem. He showed that $6m$ colors are sufficient in this more general case and that, for $m = 2$, there is a map of twelve two-region countries that requires $6m = 12$ colors. For the upper bound, Heawood argued that the average number of regions with which a region shares a border is slightly less than six. When $m = 2$, for example, the average number of regions with which a two-region pair shares a border is thus twice this, or slightly less than twelve. A simple algorithm to color with twelve colors follows from this reasoning. Martin Gardner (1997) gave an example by Scott Kim that requires twelve colors. My drawing (Figure T1) of Kim's example

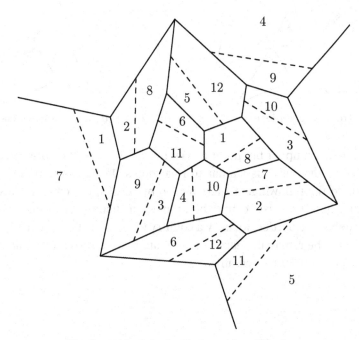

T1: Scott Kim's 2-pire that requires 12 colors

31

emphasizes its dodecahedral structure. Solid lines bound the faces of the dodeca-hedron, and a dashed line splits each face into two regions. I have labeled each in a pair of regions for a country with a number from 1 to 12.

Gerhard Ringel (1959) proposed the related "earth-moon" problem. In this version, $m = 2$ and thus one region of each country lies on the earth and the other lies on the moon. Clearly, Heawood's upper bound of twelve colors still applies, but the additional constraint suggests that fewer colors might suffice. Ringel gave an earth-moon map that requires eight colors. As reported by Martin Gardner (*SciAm*, 1980a), Thom Sulanke found an earth-moon map that needs nine colors. Whether there is any earth-moon map that needs ten, eleven, or twelve colors is still an open question.

For hinged dissections, I can do a bit better for an upper bound. The hinges in an n-piece assemblage force $n - 1$ pairs of regions in corresponding earth-moon maps to be adjacent both on the earth and on the moon. A forced adjacency in Figure T3 replaces the hinge between pieces 1 and 2 in Figure T2. Two adjacencies then replace the hinge between pieces 2, 3, and 4. I perform these replacements on both the earth and the moon maps. It follows that the average number of *countries* with which a region shares its borders is less than eleven owing to the $n - 1$ pairs of regions having adjacencies both on earth and the moon.

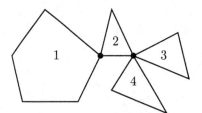

T2: From hinged pieces …

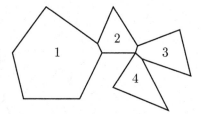

T3: … to forced adjacencies

Thus I obtain an upper bound of eleven on the number of colors needed. The method to color the pieces is similar to Heawood's: In the corresponding earth-moon maps, do the following. If there are no more than eleven countries, assign colors arbitrarily. Otherwise, count the number of distinct countries with which each country shares a border. Identify a country with at most ten such adjacencies, remove it, color the remaining earth-moon map, restore the country, and color it with a color not used for its neighbors.

CHAPTER 4

Making Squares Dance

Squares have danced together in beautiful patterns for a long, long time. Tessellations of squares graced the Islamic civilization that flourished at the turn of the first millennium. Two astronomer-mathematicians of this period, Thābit and Abū'l-Wafā, described dissections in which the pieces danced from several squares to a larger square. Dissections of squares illustrating the Pythagorean theorem danced from the Arabic manuscripts into the mathematics texts of seventeenth- and eighteenth-century Europe. And in the late nineteenth century, men such as London stockbroker Henry Perigal (1873), Belgian chevalier Paul Busschop (1876), and the Frenchman de Coatpont (1877) gave squares new movement.

Now, at the turn of the second millennium, we animate squares once again. On several websites, pieces waltz from two squares to a larger one. In this book I shall add my own bit of choreography, making squares dance with hinged motion. My first technique uses the tessellations of a thousand years ago. I superpose them as first described by the German doctoral student Paul Mahlo (1908) but constrain them to have symmetry points like those we have seen in Chapter 3. Here we will see their power.

The first example comes from the beautiful 5-piece dissection of two unequal squares to one, described by Henry Perigal (1873). The dissection, possessing 4-fold rotational symmetry, is in Figure 4.1. Although the dissection is hingeable, Perigal did not identify it as such. In the late 1980s, David Singmaster, the historian of

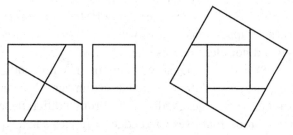

4.1: Perigal's hingeable two unequal squares to one

33

recreational mathematics, noted that the dissection is hingeable and constructed a hinged wooden model of it.

We derive Perigal's dissection by superposing tessellations as in Figure 4.2. Each symmetry point is a point of 2-fold rotational symmetry, identified by a dot. The hinged pieces are in Figure 4.3. The dissection is hinge-snug and grain-preserving. We can generalize Perigal's dissection to give a 6-piece hingeable dissection of two unequal squares to two different squares by superposing a tessellation based on the first pair of squares with a tessellation based on the second pair of squares.

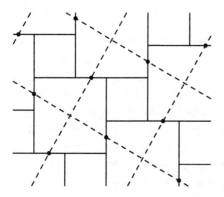

4.2: Superposition: two squares   **4.3:** Hinges for two unequal squares to one

Perigal's dissection is one from an infinite family of 5-piece dissections, the remainder of which we can find by translating (but not rotating) the two superposed tessellations relative to each other. Interestingly, Perigal's dissection seems to be the only one in this family that is hingeable. But there is another 5-piece hingeable dissection of two unequal squares to one, as we shall now see.

The other hingeable dissection is based on Thābit's 5-piece dissection, which appears in his *Risāla fi'l-hujja al-mansūba ilā Suqrāt fi'l-murabba wa qutrihi* (*Treatise on the Proof Attributed to Socrates on the Square and its Diagonals*). See (Sayili 1960). The large square contains two regions that are right triangles with hypotenuses equal to the side of the large square. For the purposes of dissection, these triangles are interchangeable, and Thābit did not indicate where each triangle should go in the large square. The hingeable variation of Thābit's dissection appears in Figure 4.4. The earliest appearance of this variation that I have found is by Yates (1940), although he did not identify it as hingeable.

Paul Mahlo (1908) and the English combinatorial mathematician Percy MacMahon (1922) superposed tessellations to derive an unhingeable arrangement of the dissection. We can derive a hingeable dissection from a different tessellation; see Figure 4.5. The tessellation that uses solid edges arranges small squares together in groups of four, and similarly for medium-size squares. Symmetry points with

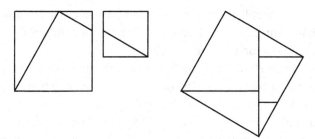

4.4: Hingeable variation of Thābit's two unequal squares to one

4-fold symmetry are at the center of such groups of squares. The other symmetry points have 2-fold rotational symmetry. The dissection differs from the normal presentation in that the pieces are arranged differently in the resulting single square. The hinged pieces swing around in Figure 4.6.

Philip Kelland (1864) identified a simplified version of Thābit's dissection as hingeable well over a century ago. This version (Figure 1.5) assumes that the two small squares are attached, so that we need just three pieces. The dissection clearly derives from the superposition of tessellations in Figure 4.5. Three different hingings are possible. Johannes Böttcher (1921) gave one (Figure 1.6), and Donald Bruyr (1963) gave another. Can you find all three?

Brodie (1884) observed that a dissection like Thābit's also applies to similar rectangles. However, I do not know how to extend the hingeable dissection in Figure 4.4 to a hingeable dissection for similar rectangles. If the length of the smaller rectangle is no longer than the width of the larger, then an appropriate modification of Perigal's dissection will work.

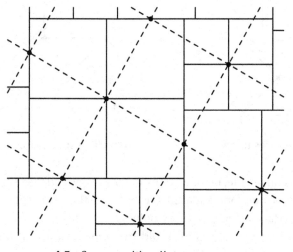

4.5: Superposition II: two squares

4.6: Hinges II

35

Other 5-piece hingeable dissections of two squares to one are possible. For example, the Belgian Luc Van den Broeck (1997) gave two different 5-piece hinged dissections for squares for $1^2 + (\sqrt{8})^2 = 3^2$.

When we arrange to have *three* squares dance together, we find an even greater variation of cases and techniques. The obvious first case is when all three squares are equal. Abū'l-Wafā gave a 9-piece dissection, which was described by the historian of mathematics Franz Woepcke (1855) and the French writer of recreational mathematics Emile Fourrey (1907). The dissection (in Figure 4.7) is rather pretty, with 4-fold rotational symmetry in the large square, and 2-fold replication and 2-fold rotational symmetry in two of the small squares.

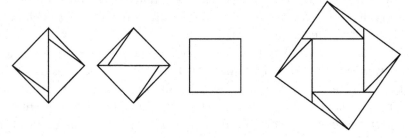

4.7: Abū'l-Wafā's three squares to one

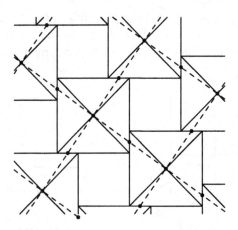

4.8: Superposition: three squares to one

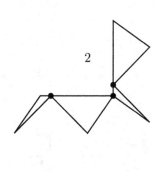

4.9: Hinges: 3 squares

Although the dissection is hingeable, no one seems to have identified it as such. We derive this dissection by forming one square from two and then superposing tessellations as in Figure 4.8. Half of the symmetry points are points of 2-fold rotational symmetry and the others are points of 4-fold rotational symmetry. The hinged pieces are in Figure 4.9.

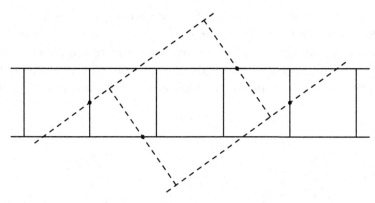

4.10: Crossposition of three squares

As charming as Abū'l-Wafā's dissection is, it does not possess the economy of more modern dissections. Paul Busschop found the 7-piece unhingeable dissection reproduced by Eugène Catalan (1873), and Monsieur de Coatpont (1877) found a 7-piece unhingeable dissection that is symmetrical but turns over two pieces. Kelland (1855) gave a dissection of a gnomon from which the 6-piece unhingeable dissection of Perigal (1891) derives. To derive an economical hinged dissection, I use the crossposition in Figure 4.10 to find the 7-piece version in Figure 4.11. I show one of three possible hingings. Trading some of the symmetry from Abū'l-Wafā's dissection for a reduction in the number of pieces, the dissection still manages to have 2-fold rotational symmetry in the large square and 2-fold replication symmetry in two of the small squares. The dissection is hinge-snug and grain-preserving. We can also derive less symmetrical 7-piece hinged dissections from Figure 3.19 and Solutions 3.2 and 11.1.

Three squares can dance together even when they are all unequal. There are 8-piece unhingeable dissections of three unequal squares to one. For the case when

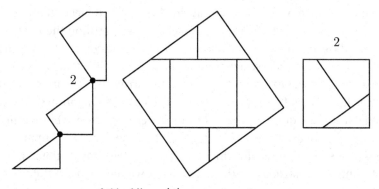

4.11: Hinged three squares to one

$x^2 + y^2 < z^2$, I modified a dissection that used two applications of Perigal's dissection. But to get a hingeable dissection, let's go back to two direct applications of Perigal's dissection, first on the x-square and y-square and then on the result of the first with the z-square. This yields the 9-piece hinged dissection in Figures 4.12 and 4.13. Since Perigal's dissection is hinge-snug and grain-preserving, this one is too.

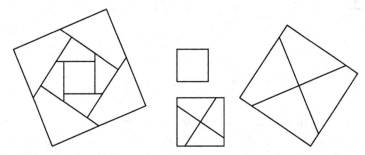

4.12: Hingeable three squares to one with $x^2 + y^2 < z^2$

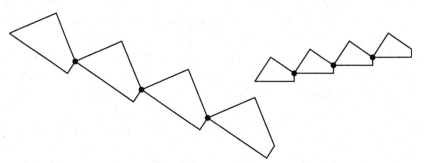

4.13: Hinged pieces for three squares to one with $x^2 + y^2 < z^2$

For the case when $x^2 + y^2 > z^2$, again there exist 8-piece unhingeable dissections. In order to obtain a hingeable dissection, we partition the w-square into three trapezoids as on the top left in Figure 4.14. Then we convert the x-square to a trapezoid in two pieces and use a continuation of the cut-line to separate the other two trapezoids. The trapezoid on the right comes from the z-square and the one on the lower left comes from the y-square. The cut-line will always intersect the bottom side of the w-square to the left of its lower right corner but within $w - x$ of that corner. There are three cases, depending on how $x + y$ compares with w. In Figure 4.14, I show the case in which $x + y > w$. We convert the z-square to a trapezoid using a TT2-strip as on the upper right; we then convert the y-square to

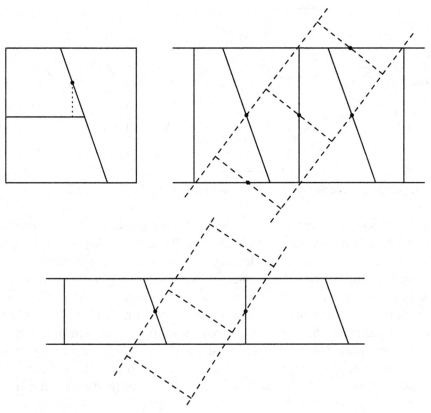

4.14: Finding hingeable three squares to one with $x^2 + y^2 > z^2$ and $x + y > w$

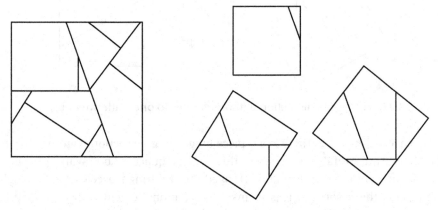

4.15: Hingeable three squares to one with $x^2 + y^2 > z^2$ and $x + y > w$

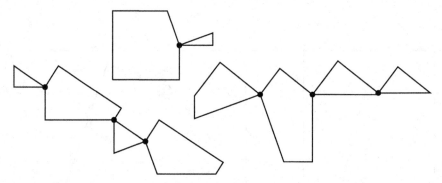

4.16: Hinged pieces for three squares to one with $x^2 + y^2 > z^2$ and $x + y > w$

a trapezoid using a PT-strip as on the lower left. The resulting 10-piece dissection does its soft-shoe in Figures 4.15 and 4.16. The dissection is hinge-snug and grain-preserving. If $x + y < w$ then we should use a TT2-strip for the y-square.

More beautiful dancing threesomes are possible when two of the squares are equal. When $x = y < z/\sqrt{2}$, there is an 8-piece hinge-snug dissection (Figure 4.17) based on the dissection of two squares to one by Perigal (1873). Just split the x- and y-squares on the diagonal and rearrange to form the smaller square in Perigal's dissection. I show the superposition of the tessellations in Figure 4.18 and the hinged pieces in Figure 4.19. The conversion to isosceles right triangles is hingeable, and the tessellations have centers of 2-fold rotational symmetry as marked.

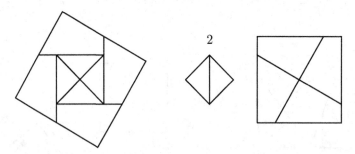

4.17: Hingeable dissection of three squares to one, with two equal I

When $x/\sqrt{2} < y = z$, there is a 9-piece hinge-snug dissection (Figure 4.20) that generalizes Abū'l-Wafā's dissection of three equal squares. Abū'l-Wafā split the y- and z-squares on the diagonal and arranged the resulting isosceles right triangles flush against the x-square. This forms a tessellation element with the repetition pattern of a square. The superposition of the tessellations is in Figure 4.21, and the hinged pieces are in Figure 4.22. The conversion to isosceles right triangles

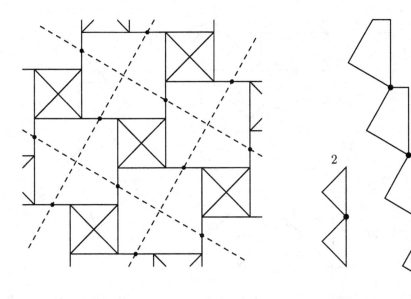

4.18: Superposition for three squares   **4.19:** Hinges: two equal

is hingeable, and the tessellations have points of both 2-fold and 4-fold rotational symmetry.

When $x = y = z/\sqrt{2}$, there is more than one 5-piece hingeable dissection.

Puzzle 4.1 Find the 5-piece hingeable dissection for $x = y = z/\sqrt{2}$ in which the z-square is not cut.

Combinations of squares with more subtle relationships lead to more challenging problems in choreography, for which we resort to slides and other special techniques. There is a 7-piece hinge-snug dissection (Figures 4.23 and 4.24) of three squares to one for the case when $x^2 + y^2 = z^2$. If we place the x-square and y-square flush together and sharing a vertex, then the line segment from the opposite corner

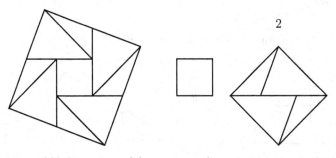

4.20: Hingeable dissection of three unequal squares to one, two equal II

41

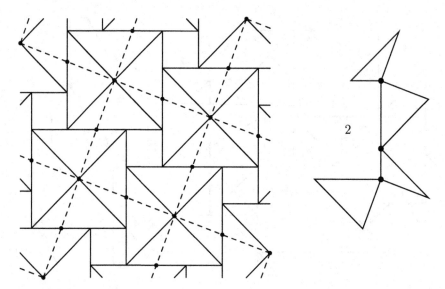

4.21: Superposition: three squares II   **4.22:** Hinges: two equal II

in the x-square to the opposite corner in the y-square is of length exactly w. This leads to a dissection of the x- and y-squares into a trapezoid that is half the area of the w-square. We can dissect the z-square to the same trapezoid and then assemble the two trapezoids to give the w-square.

Another special case is when $x + y = w$. There is a 6-piece unhingeable dissection of three squares to one for the case when $x + y = w$ and $z < 4x$. Turning to hingeable dissections, I have found a 7-piece dissection for the case when

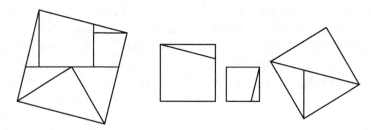

4.23: Hingeable three squares to one with $x^2 + y^2 = z^2$

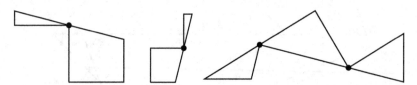

4.24: Hinged pieces for three squares to one with $x^2 + y^2 = z^2$

$1 < y/x < 2$ (and thus $y < z$). I leave the x- and y-squares uncut, transform the z-square to an $(x \times 2y)$-rectangle, and then cut the rectangle into two $(x \times y)$-rectangles. Then I use a T-slide to transform the z-square to get the dissection in Figure 4.25. The approach of forming a large square from two smaller squares and two rectangles is a technique that W. T. Tutte, C. A. B. Smith, A. H. Stone, and R. L. Brooks identified for finding squared squares, as discussed by Tutte in (Gardner 1961).

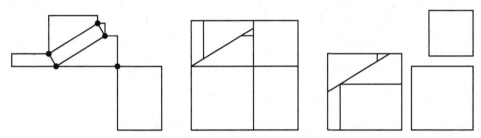

4.25: Hinged three squares to one with $x + y = w$

Since we have been so successful with squares, let's try animating their cousins, Greek Crosses. Even though there was previously no general dissection – hingeable or unhingeable – of two unequal Greek Crosses to one, I have found a hingeable one. Previously, I had found 10-piece unhingeable dissections for all Greek Crosses

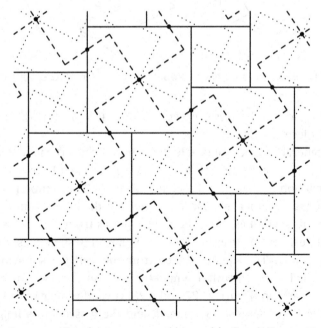

4.26: Superposition for two unequal hingeable Greek Crosses to one

43

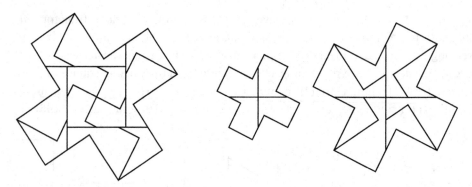

4.27: Two unequal hingeable Greek Crosses to one

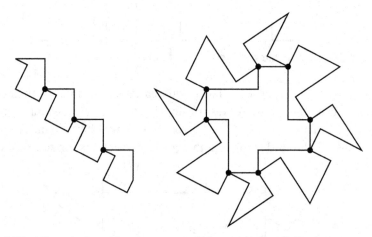

4.28: Hinged pieces for two unequal hingeable Greek Crosses to one

in Pythagoras's and Plato's classes except for $3^2 + 4^2 = 5^2$, for which Robert Reid had found a 7-piece dissection.

To get a general 12-piece hingeable dissection, I convert the x-{G} and y-{G} to squares by using Loyd's dissection in Figure 10.1. Then I arrange the squares as in the tessellation for Perigal's dissection (Figure 4.2) and superpose a tessellation of Greek Crosses rather than large squares. For the tessellation in Figure 4.26, the squares are in solid lines, the cuts in the squares from the x-{G}s and y-{G}s are in dotted lines, and the tessellation of z-{G}s is in dashed lines. Dots indicate the symmetry points. The square formed from the y-{G} has a symmetry point in its center, which I match with a symmetry point in the tessellation of z-{G}s. The resulting dissection (Figures 4.27 and 4.28) is valid whenever $4/3 \le y/x$. The pieces from the y-{G} are cyclicly hinged, and the dissection is hinge-snug and grain-preserving.

For Greek Crosses in the ratio of $1:2$, we can do better. I give a 9-piece hinge-able dissection in Figure 4.29. The dissection uses a 5-piece dissection by Henry Dudeney (*Strand*, 1926a) of a 1-{G} and a $2\sqrt{5}$-square to a $\sqrt{5}$-{G}. As luck would have it, Dudeney's dissection is cyclicly hingeable. Converting the square to a Greek Cross takes four more pieces, using a standard dissection of a square to a Greek Cross. The hinged pieces for the 2-{G} follow in Figure 4.30, with the cross-shaped area in the center being of course a hole inside of the four cyclicly hinged pieces. Alternatively, we could have hinged the four triangles at the other acute angle.

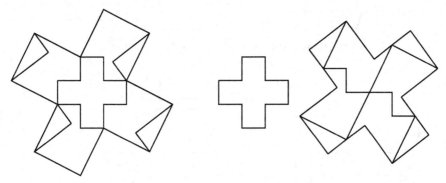

4.29: Hingeable Greek Crosses for $1^2 + 2^2 = (\sqrt{5})^2$

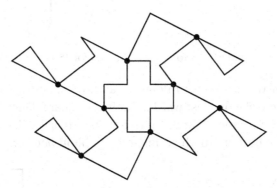

4.30: Hinged pieces for Greek Crosses for $1^2 + 2^2 = (\sqrt{5})^2$

We can also consider the case of Greek Crosses for $1^2 + 3^2 = (\sqrt{10})^2$. A diagonal of the 1-{G} equals a side of the large cross, and the diagonal of 3-{G} corresponds to another distance in the large cross. To achieve the 8-piece hinge-snug dissection in Figure 4.31, swap around small quadrilaterals like the two in the 3-{G}. Nibbling the rightmost quadrilateral out of a piece of the 3-{G} forces a trade of the quadrilateral among pieces in the 1-{G}, which leaves a hole for the other quadrilateral of the 3-{G} to fit into. The hinged pieces swing freely in Figure 4.32.

45

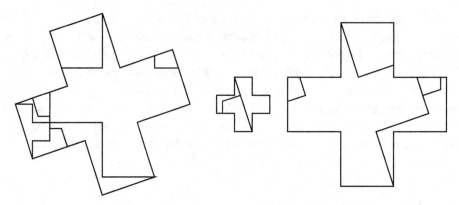

4.31: Hingeable Greek Crosses for $1^2 + 3^2 = (\sqrt{10})^2$

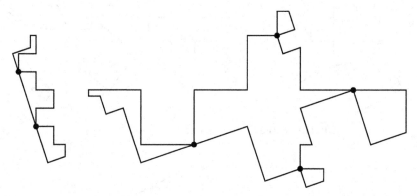

4.32: Hinged pieces for Greek Crosses for $1^2 + 3^2 = (\sqrt{10})^2$

Having taught Greek Crosses to dance, can now we swing *their* cousins, the Latin Crosses? We make one last whirl around our square dance floor, taking advantage of the special ratio of 1 : 2. I give a 9-piece hinged dissection in Figures 4.33 and 4.34. The best approach seems to be to fill in one end of an arm of the

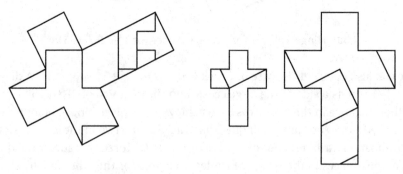

4.33: Hingeable Latin Crosses for $1^2 + 2^2 = (\sqrt{5})^2$

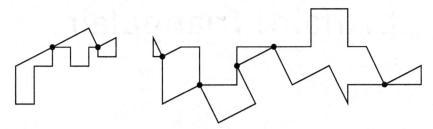

4.34: Hinged pieces for Latin Crosses for $1^2 + 2^2 = (\sqrt{5})^2$

large cross using the small cross. The longer arm seems to be the one to handle in this way.

Well, you squares, this completes your first turn on the dance floor. And even though you have turned a head or two, serious competition is up next from figures with fewer constraints.

CHAPTER 5

L'Affaire Triangulaire

In the previous chapter, we explored all manner of hinged dissections involving squares: from general illustrations of the Pythagorean theorem to various combinations of three squares, and even to dissections of crosses formed from squares. Although making squares dance is clean, wholesome fun, some readers may lust after a racier, more daring experience. Let's approach from an angle a bit less "righteous," turning away from the straight and perpendicular. Dare we experiment with triangles? And once we have taken that leap, how can we resist the temptation of hexagons and hexagrams?

Let's start by watching two triangles embrace. Although MIT professor Harry Bradley (1930) could manage only an unswinging dissection of two unequal triangles to one, Harry Lindgren (1956) found one that really did swing. He cut the small triangle from the top of the large triangle and used a Q-swing to convert the trapezoidal base to the third triangle. The 5-piece hinged dissection is in Figure 5.1. Because of the Q-swing, the dissection is hinge-snug and grain-preserving.

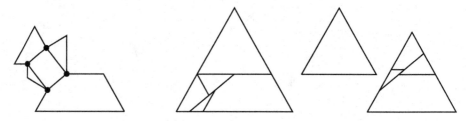

5.1: Lindgren's two unequal triangles to one

There is a second way to produce a 5-piece hinged dissection of two unequal triangles to one. If we follow the same idea as in Lindgren's dissection but use a T(a)-swing in place of a Q-swing, we get Figure 5.2. As with the previous dissection, this one is hinge-snug and grain-preserving. Due to the limitation on the range of heights for a T(a)-swing, the medium-sized triangle must have a side length at least 4/3 times that of the small triangle.

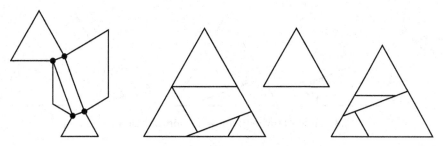

5.2: Another hinged dissection of two unequal triangles to one

Hey, you've heard about swapping partners? Let's swap triangles. We can generalize Lindgren's dissection of two unequal triangles to one to a dissection of two unequal triangles to two different ones. Cut the smaller triangles off the top of the larger triangles and convert one trapezoidal base to the other trapezoidal base using a Q-swing. The resulting hinged dissection (Figure 5.3) is hinge-snug and grain-preserving. For certain combinations of sidelengths, I give 5-piece hingeable dissections in Chapter 8.

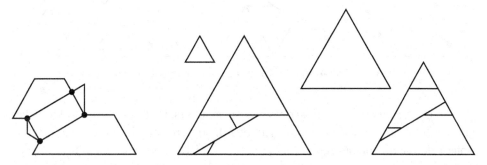

5.3: Two unequal triangles to two

Dissecting three unequal triangles to one is relatively easy: First we use Lindgren's two-into-one dissection (Figure 5.1) on the smallest two. Then we use it again, with the remaining triangle forming the trapezoidal base. This 9-piece hingeable dissection is too simple to illustrate – try drawing it yourself. When $x \leq y \leq z$ and $y \leq 2x$, we can save a piece with the 8-piece dissection in Figure 5.4. It is hinge-snug and grain-preserving. I use a variation of a Q-swing on the z-triangle and a T-strip to convert the y-triangle to a trapezoid. The latter is just a variation of our old friend, the triangle-to-square dissection.

Let's consider some special cases that give us even greater satisfaction. In Frederickson (1997) I found a 6-piece hingeable dissection of three triangles to one, where two of the triangles are equal. (Actually, I claimed to give two different dissections for different ranges of side length of the equal triangles to side length of the

5.4: Hinged three triangles to one, all unequal

third triangle. Unfortunately, the two dissections are the same; one is just a rotation of the other!) The hinged dissection, based on a Q-swing, is in Figure 5.5. It, too, is hinge-snug and grain-preserving. The dissection works whenever the side lengths of the two equal sides are smaller than twice the side length of the third triangle.

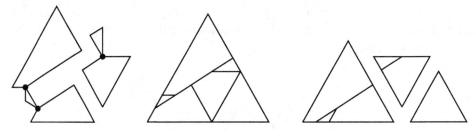

5.5: Hinged three triangles to one, with two equal

If these triangles have aroused your basic instincts, then you are bound to have a fatal attraction to our hexagon dissections. Lindgren (1964b) found a very pretty 8-piece dissection of hexagons for $x^2 + y^2 = z^2$ for the case when $y/x > \sqrt{3}$. Lindgren based it on the superposition of tessellations in Figure 5.6, which has symmetry points as indicated by dots. Lindgren's dissection comes close to being hingeable, but it splits the x-hexagon into four pieces that form two equilateral triangles that are not connected.

I have adapted his dissection to get the 10-piece hinged dissection in Figures 5.7 and 5.8. I use a 3-piece hinged gadget to reposition one triangular hole to be next to the other in the z-hexagon. The gadget consists of the two small equilateral triangles and the trapezoid in the upper right portion of the y-hexagon. The small triangles, as well as the trapezoid, are exactly half the height of the triangular hole. I fill in the hole in the z-hexagon with one of the small triangles plus a portion of the trapezoid. Sharp-eyed readers will see the T(b)-swing at work here.

Then I cut the x-hexagon into three hinged pieces to fill the combined hole at the top of the z-hexagon. The hinge-snug dissection of the x-hexagon to a 60°-rhombus is by Donald Bruyr (1963).

50

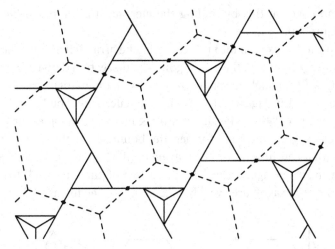

5.6: Superposed tessellations for two unequal hexagons to one

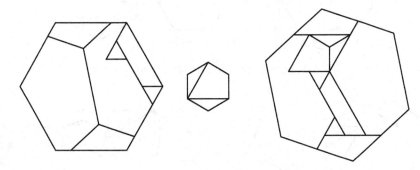

5.7: Two unequal hexagons to one

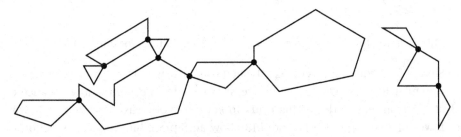

5.8: Hinged pieces for two unequal hexagons to one

For the case when $1 \leq y/x < \sqrt{3}$, Lindgren (1964b) found a seductive 9-piece dissection of hexagons for $x^2 + y^2 = z^2$. I was not able to adapt his dissection to get a hingeable dissection but instead found a 12-piece hingeable dissection by applying the T-strip technique to each of the x- and y-hexagons. However, Anton

51

Hanegraaf improved on this by finding the elegant 11-piece hingeable dissection floating in Figure 5.9.

Anton cut the y-hexagon into pieces, which he arranged around the outside of the uncut x-hexagon to form the z-hexagon. The dissection derives from a 13-piece dissection of two (unequal) hexagons to one by mathematics writer David Wells (1975), as suggested in Figure 5.10. Taking his cue from Abū'l-Wafā's dissection in Figure 4.7, Wells cut the y-hexagon into six isosceles triangles, arranged them around the x-hexagon, and then formed the boundary of the z-hexagon (dashed edges). Hanegraaf saw how to use the cutting implied by the dashed edges to hinge the pieces in the y-hexagon. He then glued two of the cuts back together. The pieces from the y-hexagon are cyclicly hinged, as we see in Figure 5.11.

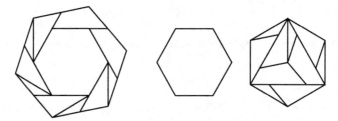

5.9: Hanegraaf's two unequal hexagons to one

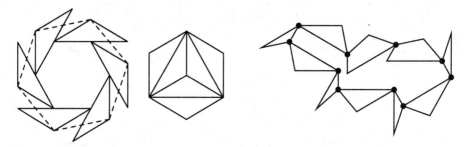

5.10: Forming the z-hexagon outline   **5.11:** Hinges for unequal hexagons

Since a hexagon consists of six attached triangles, we can easily draw hexagons on triangular lattices. On such a lattice, it is easy to find points at a distance of $\sqrt{3}$ or $\sqrt{7}$ from each other. This leads to several exotic cases for hexagons that use fewer pieces. Figures 5.12 and 5.13 show an 8-piece hinge-snug dissection of hexagons for $(\sqrt{3})^2 + 2^2 = (\sqrt{7})^2$. The cuts in the $\sqrt{3}$-hexagon give two sides of the $\sqrt{7}$-hexagon while at the same time allowing all sides in the $\sqrt{3}$-hexagon to be folded against each other.

Figures 5.14 and 5.15 show a 7-piece hinge-snug dissection of hexagons for $(\sqrt{7})^2 + 3^2 = 4^2$, just one piece more than the unhingeable dissection I had previously found. The cuts in the $\sqrt{7}$-hexagon allow the sides of length $\sqrt{7}$ to be folded against

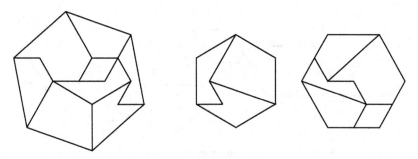

5.12: Hexagons for $(\sqrt{3})^2 + 2^2 = (\sqrt{7})^2$

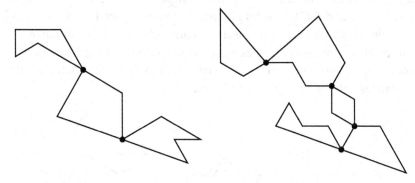

5.13: Hinged pieces for hexagons for $(\sqrt{3})^2 + 2^2 = (\sqrt{7})^2$

each other, using three large pieces. The cuts in the 3-hexagon give three sides of the 4-hexagon. A small rhombus from the $\sqrt{7}$-hexagon fills in the remaining small area in the 4-hexagon.

Some hexagon dissections have even ventured into respectability, forming a family that I will discuss in Chapter 9. Hexagrams fit into a related family, but there is one dissection that seems to do better outside the confines of domesticated bliss. Peruvian-English businessman and math buff Robert Reid found a 9-piece unhingeable dissection of two unequal {6/2}s to one for the case when the small {6/2} is

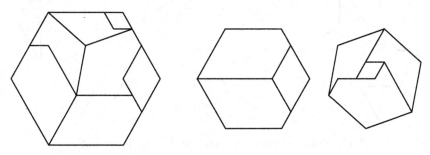

5.14: Hexagons for $(\sqrt{7})^2 + 3^2 = 4^2$

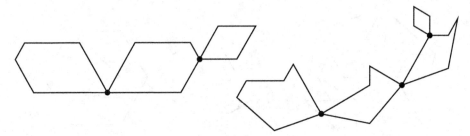

5.15: Hinges for hexagons for $(\sqrt{7})^2 + 3^2 = 4^2$

one third of the area of the larger of the two. I have found a 12-piece hinged dissection (Figures 5.16 and 5.17). The cuts in the $\sqrt{3}$-hexagram are the same as in Reid's except for one additional cut – necessary to allow all of its pieces to hinge.

Well, do those hexagon and hexagram swingers boost your heart rate? Their different sizes do help to create some tension and excitement. We'll calm down a bit when we visit the equal figures in the next chapter.

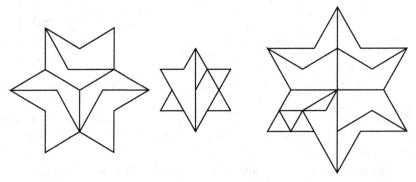

5.16: Hexagrams for $1^2 + (\sqrt{3})^2 = 2^2$

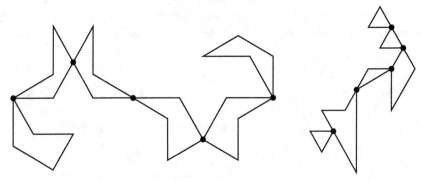

5.17: Hinged pieces for hexagrams for $1^2 + (\sqrt{3})^2 = 2^2$

Curious Case, part 2: Five Easy Pieces?

Did Henry Dudeney discover the 4-piece dissection of an equilateral triangle to a square that he displayed in his *Weekly Dispatch* column of May 4? And were the 5-piece solutions as easy as he implied in his column of April 20?

Harry Lindgren (1951) first described the T-strip method for deriving the 4-piece solution (see Figure 3.6). He crossposed a strip of squares with a strip of triangles, making sure that a symmetry point of one strip coincided with a symmetry point of the other. Dudeney did not explain any method for arriving at the 4-piece dissection, observing only that if the three angles of 60° are made to meet at a point then they will form a straight angle. This sounds a bit weak for someone who supposedly discovered what was equivalent to an application of the T-strip technique.

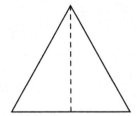 

C1: P-slide for a 5-piece dissection

On the other hand, Dudeney did explain one method clearly in his column of April 20; it led to the 5-piece dissection in Figure C1. He cut the triangle along its altitude, flipped over one of the two pieces, fit them together to form a rectangle, and then performed a P-slide. The combination of cutting and flipping to form a rectangle to which one can apply a further operation seems easy only in hindsight. He alluded to a related 5-piece dissection that involved no turning over. Perhaps that one is the dissection in Figure C2, given by Victor Klee and Stan Wagon (1991).

Another approach is to form a parallelogram by cutting a triangle of a quarter of the area off the top of the given triangle, and flipping it around to give a P-strip

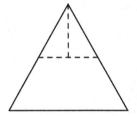 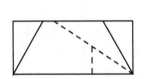

C2: A second P-slide for a 5-piece dissection

element. Then we can produce a dissection by crossposing P-strips. Again, this is a combination of techniques, with the latter being even less familiar than the P-slide. Figure C3 gives two of many possible crosspositions, each consistent with a moderately difficult problem.

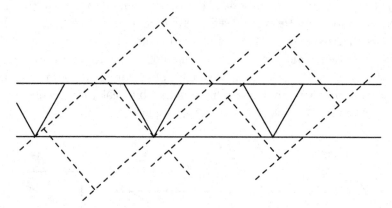

C3: Two crosspositions for 5-piece dissections

This approach allows for an infinite number of solutions in 5 pieces, depending on how we position the square on top of the parallelogram. The one in Figure C4 leads to a 4-piece dissection because we can attach two of the pieces together. However, Henry Dudeney did not give a related 5-piece dissection and thus could not show how to derive the 4-piece dissection from it.

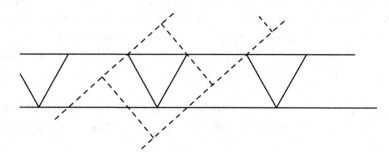

C4: Deriving the 4-piece dissection

CHAPTER 6

Return to Polygonia

In the tenth century, the Arabian mathematician and astronomer Abū'l-Wafā described a method for dissecting many equal squares to one. It works whenever the number of equal squares is $a^2 + b^2$, where a and b are integers. Alpay Özdural (2000) pointed out that this method was of considerable practical value to artisans and was apparently the result of Abū'l-Wafā's participation in meetings with these artisans. Islamic ornamentation often used geometric motifs, and mathematicians were essential in teaching practical geometry to the artisans. The artisans would have had a supply of equal squares with which to tile large square areas.

We can produce Abū'l-Wafā's dissection by superposing two tessellations of squares. The dissection of five squares to one, with $a = 1$ and $b = 2$, is an excellent example. As a bonus, the simple 9-piece dissection turns out to be hingeable and also hinge-snug. Abū'l-Wafā's method often produces too many pieces when used for larger numbers of squares. For instance, when $a = 2$ and $b = 3$, Robert Reid gave a 20-piece unhingeable dissection of thirteen squares to one, whereas Abū'l-Wafā's method leads to a 21-piece unhingeable dissection. However, Abū'l-Wafā's

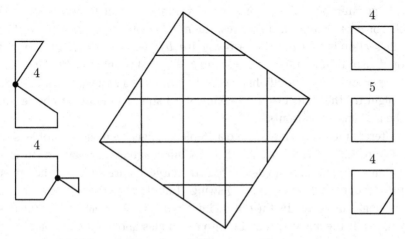

6.1: Hinged dissection of thirteen equal squares to one

method fares better when we look for hingeable dissections, since his dissection for thirteen squares to one (Figure 6.1) is hingeable, whereas Reid's is not. Abū'l-Wafā's method also works well for seventeen squares to one.

Puzzle 6.1   For seventeen squares to one, Abū'l-Wafā's method also gives a hingeable dissection. Find the set of hinged pieces.

Concerning hingeable dissections of squares, I have not had the opportunity to participate in meetings with artisans – Does anyone want hinged mosaics on their floors? – but I am open to any appropriate collaboration. For ten squares to one with $a = 1$ and $b = 3$, Abū'l-Wafā's method gives a 16-piece unhingeable dissection. At the expense of two more pieces and a bit more work, we can find the 18-piece hinged dissection in Figure 6.2.

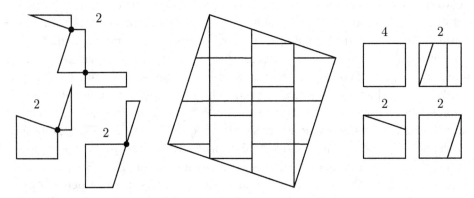

6.2: Hinged dissection of ten equal squares to one

For larger numbers of squares, artisans may not want to adapt Abū'l-Wafā's method. For $a^2 + b^2$ identical squares, the number of pieces for Abū'l-Wafā's unhinged dissection is $a^2 + b^2 + 2(a + b + \gcd(a, b))$, where $\gcd(a, b)$ is the greatest common divisor of a and b. For $a = 3$ and $b = 5$, that number is 48. Also, it is not readily apparent how to modify his method to produce a hingeable dissection. The reader might try the same example – namely, 34 squares to one – to better understand the artisans' constraints.

By different methods, I have found a 46-piece hingeable dissection of 34 squares to one (Figure 6.3). Since I leave 25 of the 34 small squares whole and cut eight of the remaining nine into rectangles, I show the large square and only the one small square that I cut using a modified Q-swing. I apply the Q-swing to an imagined larger rectangle, indicated by the dotted lines. To maneuver pieces into the desired positions, I split the squares A and B into rectangles and rotate one piece each by 90°. For the other squares that I cut, pieces will rotate by 180°. Take note, artisans,

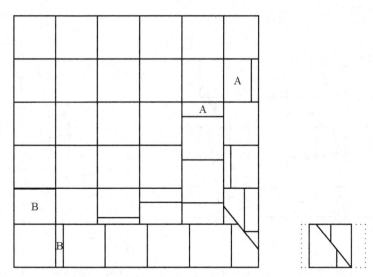

6.3: Hingeable dissection of 34 equal squares to one

that the thick-looking line above the fat piece labeled with a B is actually a rather thin rectangle.

Branching out from squares, we can apply an idea similar to Abū'l-Wafā's to Greek Crosses. Henry Dudeney (*Dispatch*, 1900a) gave the 5-piece hingeable dissection of two Greek Crosses to one in Figure 6.4. Harry Lindgren (1964b) showed how to produce it by superposing tessellations as in Figure 6.5. The hinged piece is in Figure 6.6. I have found no evidence that Dudeney knew that the dissection is hingeable. The dissection is also grain-preserving.

Lindgren (1964b) gave a 12-piece unhingeable dissection of five Greek Crosses to one. Lindgren clustered four of the small crosses in the center of the large one and then distributed quarters of the fifth one to the ends of the arms of the large cross. To get a hingeable dissection, we center the dissection on one of the five

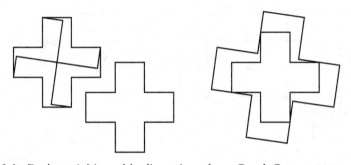

6.4: Dudeney's hingeable dissection of two Greek Crosses to one

59

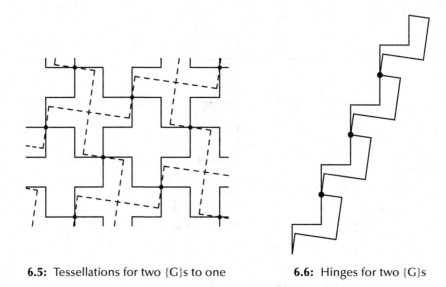

6.5: Tessellations for two {G}s to one   **6.6:** Hinges for two {G}s

small crosses. Figure 6.7 shows the tessellation elements for small crosses and the large one. Dots identify the symmetry points. Four of these points appear in the middle of cross edges and force 2-fold rotational symmetry; the other four appear at vertices and force 4-fold symmetry. The resulting 13-piece hinge-snug dissection stands in Figure 6.9, and the hinged pieces swing in Figure 6.8.

There is another hinging of the same set of pieces that is grain-preserving. In fact, one of its hinges is frozen, which perhaps accounts for its not falling on a symmetry point.

Puzzle 6.2   Find a hinging of Figure 6.9 that is grain-preserving.

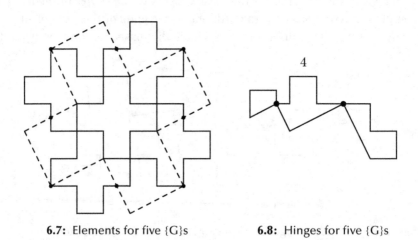

6.7: Elements for five {G}s   **6.8:** Hinges for five {G}s

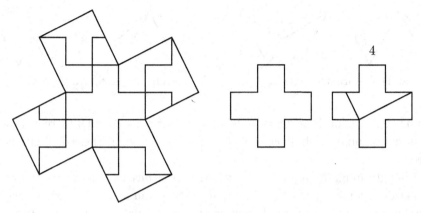

6.9: Five Greek Crosses to one

In a fashion similar to Abū'l-Wafā's, Lindgren (1964b) gave an approach for dissecting triangles. Lindgren based it on tessellations of triangles and gave economical dissections of $a^2 + ab + b^2$ triangles to one. The side of the large triangle corresponds to the side opposite a 120° angle in a triangle whose other sides have length a and b. The Australian engineer David Paterson showed in (1989) that the number of pieces is $a^2 + ab + b^2 + 3(a + b - \gcd(a, b))$.

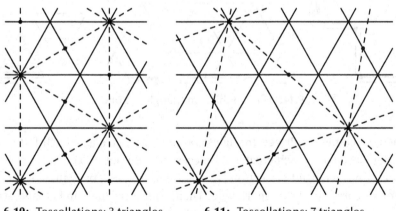

6.10: Tessellations: 3 triangles   **6.11:** Tessellations: 7 triangles

This method sometimes produces hingeable dissections. There is a 6-piece dissection of three triangles to one that Plato described in *Timaeus*. The superposed tessellations are in Figure 6.10, with dots indicating the symmetry points. Taking $a = 1$ and $b = 2$, there is a 13-piece dissection of seven triangles to one given by Polish mathematician Hugo Steinhaus (1960). We can see the superposition of the tessellations in Figure 6.11; again, dots indicate the symmetry points.

61

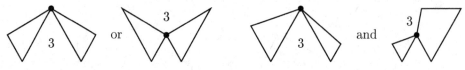

6.12: Hinges for 3 triangles to 1   **6.13:** Hinges for 7 triangles to 1

The hinged pieces for these dissections follow in Figures 6.12 and 6.13, respectively. The alternative on the right in Figure 6.12 is grain-preserving. These dissections are hinge-snug.

Although Lindgren's approach for triangles is as nice as Abū'l-Wafā's approach for squares, it suffers from the same deficiency: it can sometimes be surpassed. David Collison found a 12-piece dissection of seven triangles to one. Unfortunately, that one is unhingeable, so the 13-piece dissection prevails in the world of hinges.

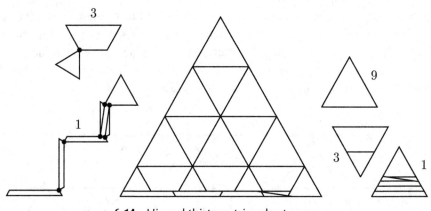

6.14: Hinged thirteen triangles to one

When we come to thirteen triangles to one, Lindgren's approach falters. The tessellation-based approach creates 22 pieces that are not hingeable. By contrast, there is a 21-piece hinged dissection (see Figure 6.14). First, pack nine of the triangles into one corner of the large triangle. Then, split three more in half and arrange them to fill in a parallelogram adjacent to the area occupied by the nine triangles. Finally, slice off the bottom two layers of the thirteenth triangle, and use a Q-swing to convert the remainder to a thin trapezoid with a small triangle perched atop. This dissection is hinge-snug and grain-preserving.

Harry Lindgren (1964b) showed how to use tessellations to find dissections of $a^2 + ab + b^2$ hexagons to one. Derived by superposing tessellations as in Figure 6.15, his dissection of three hexagons to one is hinge-snug; the hinged pieces open up in Figure 6.17.

62

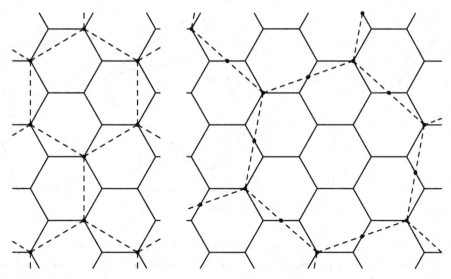

6.15: Tessellations: 3 hexagons   **6.16:** Tessellations: 7 hexagons

The case of seven hexagons to one is more interesting. Lindgren gave a 12-piece dissection that is, unfortunately, unhingeable. By shifting the relative positions of the superposed tessellations as in Figure 6.16, we can get a 13-piece hinge-snug dissection. Dots indicate points of 3-fold rotational symmetry in the superposition, leading to the hinged pieces on the left of Figure 6.18. Interestingly, the tessellation elements also possess 2-fold rotational symmetry centered at the midpoints of the sides of the large hexagons. Focusing on these, we can obtain the alternate set of hinged pieces shown on the right in Figure 6.18. The dissection based on this latter set is grain-preserving.

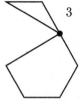

6.17: Hinges: 3 to 1

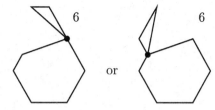

or

6.18: Hinges: 7 hexagons to 1

We come at last to the case of six pentagons to one. In (1989), Alfred Varsady, a Hungarian technical designer who lives in Germany, found a clever 21-piece un-hingeable dissection that we can derive using polyhedral tessellations. Robert Reid found a 19-piece dissection that was not based on tessellations, which I modified to

63

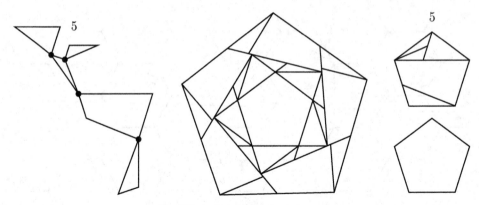

6.19: Hinged six pentagons to one

save a piece (by turning over two pieces) and which Anton Hanegraaf modified to avoid turning over any pieces. Yet once again, requiring hingeability seems to give new life to an old dissection. I have modified the approach in Alfred's dissection to give a 26-piece hinged dissection (Figure 6.19). Will this inspire my artisans?

CHAPTER 7

Who's the Squarest of Them All?

We have already made squares dance by finding hingeable dissections of two squares to one, two squares to two others, and three squares to one. We have identified a number of classes for which we need fewer pieces, mirroring our experience with unhingeable dissections. There are even better dissections when the side lengths of the squares satisfy certain integer identities. These identities form a number of classes. Do similar classes exist for hingeable dissections of squares? Yes, there are such classes, in addition to some isolated cases that require fewer pieces. Now let's enjoy these rational dissections in all their glorious squareness.

For two squares to one, we already know two general techniques from Chapter 4 that give us 5-piece hingeable dissections, so we are boxed in to finding 4-piece dissections. I know of only one integer identity, namely $3^2 + 4^2 = 5^2$, for which a 4-piece hingeable dissection exists. Sam Loyd and Henry E. Dudeney gave a number of 4-piece unhingeable dissections of squares for $3^2 + 4^2 = 5^2$. I show my hinge-snug dissection in Figure 7.1.

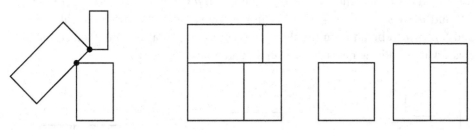

7.1: Hinged dissection of squares for $3^2 + 4^2 = 5^2$

When dissecting a pair of unequal squares to a different pair, we find ourselves similarly boxed in: We must look for 5-piece dissections, because we know of a general technique that produces 6-piece dissections. Robert Reid found a method for producing 5-piece unhingeable dissections for squares in the double-square difference class. One member of this class is the identity $13^2 + 18^2 = 22^2 + 3^2$, for which I have found the 5-piece hinged dissection in Figure 7.2. I base my dissection loosely

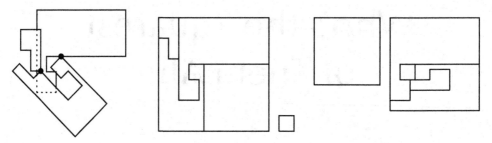

7.2: Hinged dissection of squares for $13^2 + 18^2 = 22^2 + 3^2$

on Reid's, but I cut portions out of pieces to accommodate other pieces that swing around to fit into them. The dissection is so wobbly hinged that at least two pieces always overlap except when the pieces form the 18-square or the 22-square. However, the dissection is hinge-snug.

Turning now to dissections of three squares to one, we can find all integral solutions to $x^2 + y^2 + z^2 = w^2$ by using a method of the nineteenth-century French mathematician V. A. Lebesgue. Choose m, n, p, q to be positive integers with $m^2 + n^2 > p^2 + q^2$ and $mq > np$. Then *Lebesgue's formula* is:

$$x = m^2 + n^2 - p^2 - q^2, \quad y = 2(mp + nq), \quad z = 2(mq - np);$$
$$w = m^2 + n^2 + p^2 + q^2.$$

I have found a class of 7-piece dissections that is related to the Pythagoras-plus class in my first book. We can express the new class in terms of Lebesgue's formula by choosing p to be an odd number and taking $n = 2$, $q = 2$, and $m = p + 2$. The first two members of our class, corresponding to $p = 1$ and $p = 3$, are $8^2 + 14^2 + 8^2 = 18^2$ and $16^2 + 38^2 + 8^2 = 42^2$. All values in the solutions are multiples of 2. If we had taken p to be an even number, we would have obtained the Pythagoras-plus class. Thus I call the new class the Pythagoras-plus-twin class.

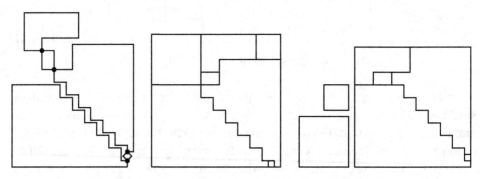

7.3: Hinged dissection of squares for $16^2 + 38^2 + 8^2 = 42^2$

Figure 7.3 illustrates the dissection for $16^2 + 38^2 + 8^2 = 42^2$. The dissections result from applying the hinged Q-step technique twice: converting the $(2(p^2 + 2) \times 2(p^2 + 4))$-rectangle in the lower right corner of the w-square to a $(2(p^2 + 4) \times 2(p^2 + 2))$-rectangle, then converting the $(2p^2 \times 4(p + 2))$-rectangle between the x- and z-squares in the w-square to a $(2p(p + 2) \times 4p)$-rectangle in the upper left corner of the y-square. The dissections are hinge-snug and grain-preserving.

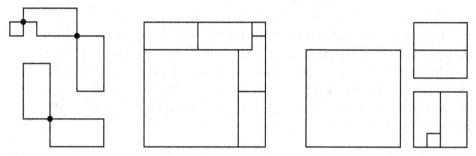

7.4: Hinged dissection of squares for $7^2 + 4^2 + 4^2 = 9^2$

There are 5-piece unhingeable dissections for all members of the PP-double class, in which $y = z$. I have not found a nice hingeable dissection for all members but do have a 6-piece hinge-snug and grain-preserving one for $7^2 + 4^2 + 4^2 = 9^2$, as shown in Figure 7.4.

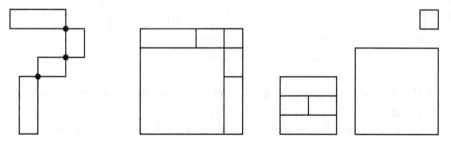

7.5: Hinged dissection of squares for $2^2 + 6^2 + 9^2 = 11^2$

Of course, a 4-piece hinged dissection for $1^2 + 2^2 + 2^2 = 3^2$ is too easy to count as a puzzle. Another identity for which there is a 5-piece unhingeable dissection is $2^2 + 6^2 + 9^2 = 11^2$; I found a 6-piece hinged dissection for this case (see Figure 7.5). The unhingeable dissection is more or less unrelated to the hinged one and is mentioned only to emphasize that we pay a 1-piece penalty to get a hinged dissection.

For $2^2 + 6^2 + 3^2 = 7^2$, Sam Loyd (*Home,* 1908a) gave a 5-piece unhingeable dissection of squares. The best I have found for a hingeable dissection is the 6-piece one

67

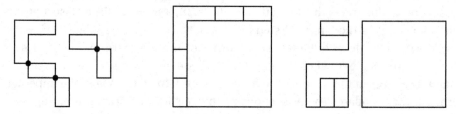

7.6: Hinged dissection of squares for $2^2 + 6^2 + 3^2 = 7^2$

shown in Figure 7.6, which is hinge-snug and grain-preserving. But this is not so bad, as Henry Dudeney (1931) had asked for a 6-piece dissection without any hinging.

There are 5-piece unhingeable dissections for all members of Cossali's class, which has $n = p = q = 1$ and m odd. For $m = 7$, this is $48^2 + 16^2 + 12^2 = 52^2$, which reduces to $12^2 + 4^2 + 3^2 = 13^2$. I have found the 7-piece hinge-snug and grain-preserving dissection in Figure 7.7.

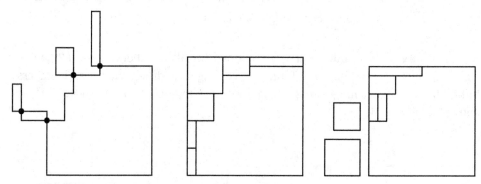

7.7: Hinged dissection of squares for $12^2 + 4^2 + 3^2 = 13^2$

In (1946), the American analyst of games, Geoffrey Mott-Smith, gave a 5-piece unhingeable dissection for squares for $1^2 + 8^2 + 4^2 = 9^2$. I have found a 6-piece one for $1^2 + 8^2 + 4^2 = 9^2$. This hinged dissection (Figure 7.8) has a lovely symmetry.

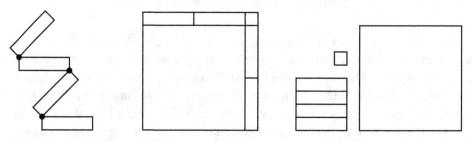

7.8: Hinged dissection of squares for $1^2 + 8^2 + 4^2 = 9^2$

Figure 7.9 displays a 6-piece hinged dissection of squares for $8^2 + 9^2 + 12^2 = 17^2$. Again, this is one piece more than we need for an unhingeable dissection. I convert the z-square into two $(x \times y)$-rectangles and fill out the w-square with these rectangles plus the x- and y-squares.

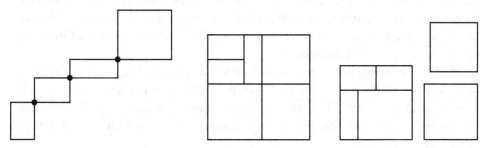

7.9: Hinged dissection of squares for $8^2 + 9^2 + 12^2 = 17^2$

From Chapter 4, we know that there is a 7-piece hingeable dissection for squares for $x^2 + y^2 + z^2 = w^2$ when $x + y = w$, $x < y$, and $y < 2x$. However, one of the pieces is a small triangle and not all cuts are parallel to the sides of the squares.

Puzzle 7.1 Find a 7-piece hingeable dissection of squares for $18^2 + 25^2 + 30^2 = 43^2$ such that all cuts are parallel to the sides of the squares.

Let's close out the chapter with a dissection that exploits a relation identified many years ago. Dudeney (*Strand*, 1923a) gave a 6-piece unhingeable dissection for $12^2 + 15^2 + 16^2 = 25^2$. There is a 7-piece hinged dissection, in Figure 7.10. Using the Q-step technique, I convert the 15-square to a (9×25)-rectangle; then I convert the 12-square to a (9×16)-rectangle. The dissection is hinge-snug.

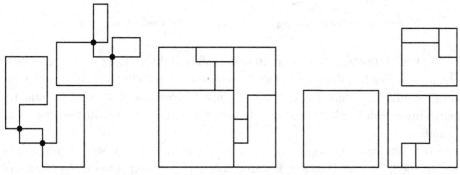

7.10: Hinged dissection of squares for $12^2 + 15^2 + 16^2 = 25^2$

Turnabout 2: Bracing Regular Polygons

In his math games column, Martin Gardner (*SciAm,* 1963a) showed Raphael Robinson's problem of bracing a square. The square consists of four rods attached at their endpoints. The four rods by themselves do not form a rigid structure because the vertices where the rods meet act like hinges. Thus, if the square is not braced, we can easily squash it into a rhombus. Our goal is to brace the square by placing, in the same plane as the square, the fewest number of additional rods of the same length as those forming the square.

The charm of this problem is that we seek to negate the hingeability that we so prize throughout the rest of this book! The rods may not cross each other and may touch only at their ends. If we allowed three dimensions then the solution would be simple, because we could then form a regular octahedron with just eight additional rods.

Gardner (*SciAm,* 1963b) stated that Robinson had found two solutions that each use 31 rods in addition to the four that form the square. Later, Gardner (*SciAm,* 1964a) reported that 44 readers had found a 25-rod solution and that seven readers – G. C. Baker, Joseph H. Engel, Kenneth J. Fawcett, Richard Jenney, Frederick R. Kling, Bernard M. Schwartz, and Glenwood Weinert – had found the 23-rod solution in Figure T4.

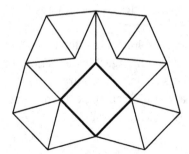 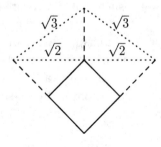

T4: 23-rod solution for bracing a square   **T5:** Underlying geometry

The key to this and other solutions is to induce right triangles of side lengths 1, $\sqrt{2}$, and $\sqrt{3}$. The length of $\sqrt{3}$ is easy to construct by forming adjacent equilateral triangles, and the length 1 is trivial. In Figure T5, solid lines identify the square, dashed lines indicate three crucial rods, and dotted lines represent the lengths of $\sqrt{2}$ and $\sqrt{3}$.

Can we brace other regular polygons in the plane? Of course, the hexagon is easy. In 1963, Thomas H. O'Beirne found solutions for the pentagon (64 rods), octagon (105 rods), decagon (183 rods), and dodecagon (45 rods). His solution for the pentagon dazzles us in Figure T6.

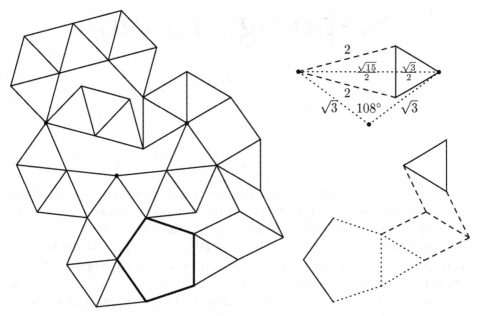

T6: O'Beirne's 64-rod solution for a pentagon   **T7:** Underlying geometry

Figure T7 illustrates two key insights. As shown on the top, O'Beirne found a nifty way to construct 108°, which is the interior angle of a regular pentagon. He constructed an isosceles triangle with two sides of length 2 (dashed edges) and one of length 1; he then attached an equilateral triangle to it (solid edges). The height of the equilateral triangle is $\sqrt{3}/2$, and the height of the isosceles triangle is $\sqrt{15}/2$, for a total of $\sqrt{3}(1 + \sqrt{5})/2$. This is the length of a diagonal in a regular pentagon with side length $\sqrt{3}$. Dots identify the vertices of a corresponding triangle in Figure T6.

O'Beirne transferred the 108° angle to the actual pentagon by using a rhombus. However, his real inspiration was in forcing all of the remaining angles of the pentagon to be the same. He attached an equilateral triangle to the rightmost side of the pentagon. The triangle's sides are parallel to the rightmost equilateral triangle of the construction only when the rightmost side of the pentagon is at the correct angle. With the two leftmost sides of the pentagon and the rightmost equilateral triangle all in fixed position (solid edges), O'Beirne used a pair of rhombuses (dashed edges) to force the edges (dotted) of the attached triangle and the remainder of the pentagon into the correct position.

Erich Friedman (2000) considered the problem of bracing the regular polygons when rods are allowed to overlap. In this case, Erich needed only seventeen additional rods to brace a square.

CHAPTER 8

Stepping Around

The step technique works for figures other than squares. The most obvious candidates are triangles, but we can also find related dissections for pentagons, hexagons, hexagrams, and Greek Crosses. So let's lift our feet and step around.

Triangles sometimes outperform squares. In the last chapter, we did not find an infinite class of solutions for $x^2 + y^2 = z^2$ that has 4-piece hingeable dissections of squares, nor an infinite class of solutions for $x^2 + y^2 = z^2 + w^2$ that has 5-piece hingeable dissections of squares. But for triangles there are such classes.

One of the simplest of the triangle dissections is a demonstration of $3^2 + 4^2 = 5^2$. Harry Bradley (1930) gave a 4-piece dissection that turns out to be hingeable. The hinged dissection (see Figure 8.1) requires just two cuts in the 3-triangle and none in the 4-triangle. The dissection is hinge-snug and grain-preserving. The latter property allows us to apply the dissection not only to equilateral triangles but also to triangles that are merely similar. The same holds for other grain-preserving dissections in this chapter.

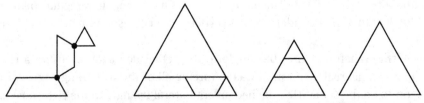

8.1: Bradley's hingeable dissection of triangles for $3^2 + 4^2 = 5^2$

More generally, let's consider triangles in Plato's class. The fourth-century Greek mathematician Diophantus gave a method for finding basic solutions to $x^2 + y^2 = z^2$, and the seventh-century Indian mathematician Brahmagupta stated it explicitly: Each solution is of the form $x = m^2 - n^2$, $y = 2mn$, and $z = m^2 + n^2$, where m and n are relatively prime and $m + n$ is an odd number. Plato's class corresponds to those solutions in Diophantus's method for which $n = 1$.

Software engineer David Collison (1979) gave 4-piece dissections for triangles in Plato's class, and these dissections are hingeable. His hinged dissection of triangles

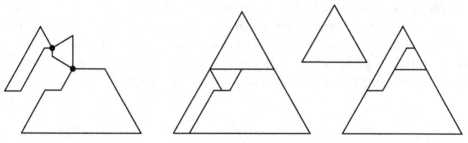

8.2: Hinged dissection of triangles for $15^2 + 8^2 = 17^2$

for $15^2 + 8^2 = 17^2$ (see Figure 8.2) is hinge-snug and grain-preserving. This dissection should not be a surprise to those who have already read Chapters 3 and 5. In Figure 5.1 we have seen how to dissect two triangles to one using a Q-swing, and in Figure 3.13 we have seen how to produce a step dissection from a Q-swing.

Let's next step into hingeable dissections of two triangles to two different triangles. Diophantus gave a specific derivation of $7^2 + 4^2 = 8^2 + 1^2$, from which we can infer a method for finding all integral solutions to $x^2 + y^2 = z^2 + w^2$. The Italian mathematician Leonardo of Pisa (Fibonacci) stated and explicitly proved the formula in 1225. Choose positive integers m, n, p, q such that $n < m$, $p < q$, and $mp \neq nq$. Then *Fibonacci's formula* is:

$$x = mp + nq, \quad y = mq - np, \quad z = mq + np, \quad w = |mp - nq|.$$

There are two classes of 5-piece dissections of triangles for $x^2 + y^2 = z^2 + w^2$: the Pythagoras-extended and the Plato-extended classes. For certain of these cases, the dissections are hingeable. In particular, the Plato-extended class applies whenever $n = 1$, and half of its members are hingeable. My method for the Plato-extended class gives hingeable dissections whenever m is even. I call this subclass the Plato-semiextended class. I restate my method as Method Plato-SE.

Method Plato-SE: For triangles in the Plato-semiextended class.

1. Cut an x-triangle from the top of the z-triangle.

2. Flip-down step starting $2p$ right of the left corner of z-triangle:
   [$m/2 - 1$ times]:
      {Move/cut up-right $2(q - p)$; Move/cut right $2p$.}
   Mark the current position.
   Move/cut up-right $q - p$; hinge at the end of the cut.

3. Starting p to the left of the marked position,
   move/cut up-left all the way through the step piece.
   Hinge at the beginning of the cut.

I have simplified Method Plato-SE to reflect the fact that m is even. The w-triangle will sit at the apex of the y-triangle precisely when the cut in step 3 extends up to the bottom edge of the x-triangle. Otherwise the w-triangle fits into a cavity on the right side of the y-triangle. As stated, $n = 1$ and m must be an even whole number, but in fact p and q need not be whole numbers.

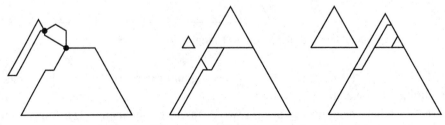

8.3: Method Plato-SE applied to $27^2 + 11^2 = 29^2 + 3^2$

In Figure 8.3, I illustrate Method Plato-SE for $m = 4$, $p = 1$, and $q = 7$, which gives $11^2 + 27^2 = 29^2 + 3^2$. In this case, the w-triangle does not sit atop the y-triangle. In Figure 8.4, I illustrate the method for $m = 6$, $p = 3$, and $q = 5$, which gives $23^2 + 27^2 = 33^2 + 13^2$. In this case, the w-triangle does sit atop the y-triangle. All dissections produced by Method Plato-SE are hinge-snug and grain-preserving.

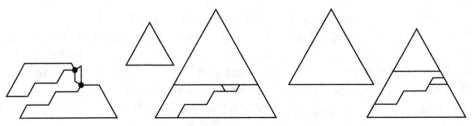

8.4: Method Plato-SE applied to $23^2 + 27^2 = 33^2 + 13^2$

Pentagons follow in the footsteps of triangles. We can slice a pentagon into three isosceles triangles, apply an appropriate hingeable dissection to each triangle, and then glue pieces back together across the the slice lines. For instance, we can find an 8-piece hinged dissection for $8^2 + 15^2 = 17^2$. Dotted lines in Figure 8.5 indicate where we have glued. The dissection is wobbly hinged, yet it is grain-preserving.

We can can apply the same technique to any $\{p\}$ for a solution in Plato's class, giving a hingeable dissection in $2p - 2$ pieces. For hexagons and $3^2 + 4^2 = 5^2$, a

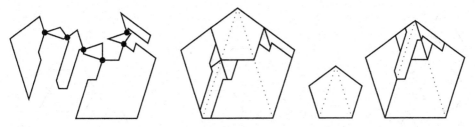

8.5: Hinged pentagons for $8^2 + 15^2 = 17^2$

special approach does better. University of Connecticut logician Jim Schmerl (1973) gave an unhingeable 5-piece dissection of hexagons for $3^2 + 4^2 = 5^2$. A hinged dissection that uses 7 pieces is shown in Figure 8.6. Following Schmerl's approach, we place an uncut 4-hexagon in one corner of the 5-hexagon. Then we cut and hinge the 3-hexagon symmetrically and wrap it around the 4-hexagon.

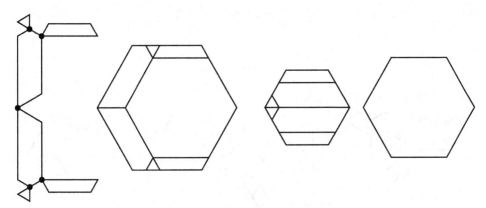

8.6: Hinged hexagons for $3^2 + 4^2 = 5^2$

A special approach also seems best for hexagons for $1^2 + 2^2 + 2^2 = 3^2$, for which there is a simple 6-piece unhingeable dissection. Robert Reid found a nifty 7-piece hinged dissection (Figure 8.7). Robert positioned the 1-hexagon in the center of the 3-hexagon; he then cut and hinged the 2-hexagons identically, swinging the smaller pieces around to help form the outer portions of the 3-hexagon. The dissection is hinge-snug and grain-preserving.

Yet another special approach works well for hexagrams for $1^2 + 2^2 + 2^2 = 3^2$. Robert found an unhingeable 9-piece dissection of two hexagrams for $1^2 + 2^2 + 2^2 = 3^2$. In my 13-piece hinged dissection (Figure 8.8), I leave the 1-hexagram uncut and place it in the center of the 3-hexagram. I then cut and hinge the 2-hexagrams

75

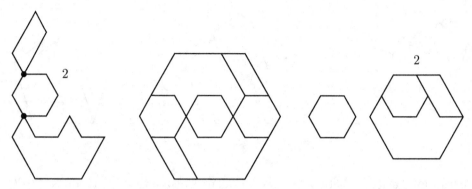

8.7: Reid's hinged dissection of hexagons for $1^2 + 2^2 + 2^2 = 3^2$

identically so that they wrap around the 1-hexagram. Unfortunately, the hinges are wobbly, but at least the dissection is hinge-snug.

The approach illustrated in Figure 8.8 works on any star of the form $\{2m/(m-1)\}$ to give a $(4m+1)$-piece dissection for $1^2 + 2^2 + 2^2 = 3^2$.

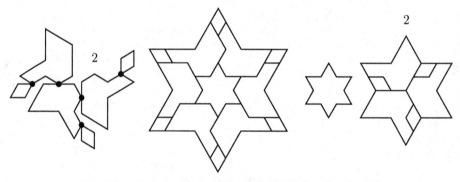

8.8: Hinged dissection of hexagrams for $1^2 + 2^2 + 2^2 = 3^2$

Puzzle 8.1 Find a 17-piece hinged dissection of $\{8/3\}$s for $1^2 + 2^2 + 2^2 = 3^2$.

We have devolved so completely into special approaches that we seem to be stepping around the step technique! Let's look at one more example, and then we are done. I have identified an unhingeable 7-piece dissection of Greek Crosses for $1^2 + 2^2 + 2^2 = 3^2$; Figure 8.9 shows my 12-piece hingeable dissection. I leave the 1-cross uncut and then cut the 2-crosses symmetrically, except for the two pieces that would overlap. We can verify from the hinged pieces in Figure 8.10 that the dissection is hinge-snug.

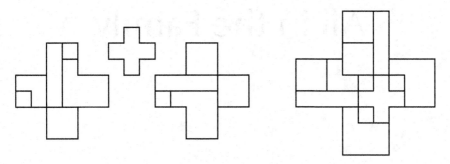

8.9: Greek Crosses for $1^2 + 2^2 + 2^2 = 3^2$

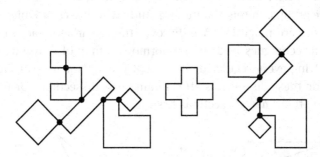

8.10: Hinged pieces for Greek Crosses for $1^2 + 2^2 + 2^2 = 3^2$

CHAPTER 9

All in the Family

Genealogical research is never more fascinating than when it brings together seemingly disparate groups. In this chapter, we study two classes of 7-piece dissections of hexagons that are hingeable. We discover the remarkable fact that variations in my hexagon techniques lead to an enormous family of polygon and star dissections, involving every regular figure except for the square and the equilateral triangle. I made these discoveries after finding new dissections for the hexagons, so let's examine those new dissections first.

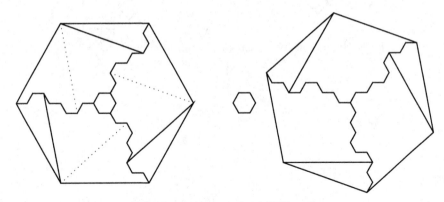

9.1: Hexagons for $1^2 + (\sqrt{63})^2 = 8^2$

One class of hexagon dissections, which I call the *Tri-minus* class, has $x = 1$ and $z = 3n^2 - 3n + 2$. The Method TM that I give here is akin to an earlier method of mine. The direction *up-right* is 60° up and to the right from the horizontal. Going down instead of up gives the direction *dn-right*. We see the example for $n = 2$ in Figure 9.1, and the hinged pieces swing out in Figure 9.2. Try applying the method when $n = 1$.

Method TM: Wobbly steps for hexagons in the Tri-minus class.

Orient the z-hexagon with two sides horizontal.

Cut a 1-hexagon with two sides horizontal out of the center.

Repeat 3 times:

  Wobbly step from left vertex:

    Move/cut right 1.

    [$n - 1$ times]: {Move/cut up-right 1; Move/cut right 1.}

    Mark the current position.

    [$n - 1$ times]:

      {[n times]: {Move/cut dn-right 1; Move/cut right 1.}

       [$n - 1$ times]: {Move/cut up-right 1; Move/cut right 1.}}

  Move/cut from the last mark to the vertex

    that is counterclockwise from the left vertex.

  Rotate the z-hexagon 120° clockwise about its center.

There is a family resemblance between this approach and my earlier one, but three of the pieces are thinner. This second approach works because six equilateral triangles form a regular hexagon. If we replace each cut of length y in the z-hexagon in Figure 9.1 by a cut from the same vertex but at an angle of 60° from the original cut, we get my original dissection. Dotted lines indicate these cuts in the z-hexagon of Figure 9.1. All dissections produced by this method are hinge-snug.

The other class of hexagon solutions, which I call the *Tri-root* class, has $x = 1$, $y = \sqrt{3n^2}$, and $z = \sqrt{3n^2 + 1}$. I give the hingeable dissection for the case $n = 3$ in

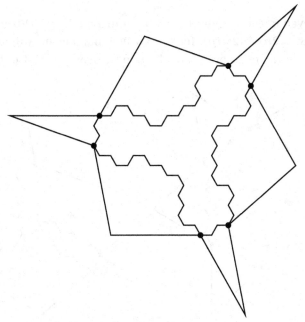

9.2: Hinges for hexagons for $1^2 + (\sqrt{63})^2 = 8^2$

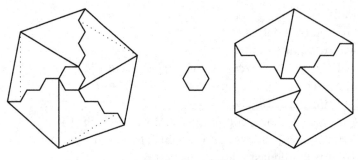

9.3: Hexagons for $1^2 + (\sqrt{27})^2 = (\sqrt{28})^2$

Figure 9.3. We can see from the hinged pieces in Figure 9.4 that the dissection is hinge-snug.

The Tri-root class of solutions also has its fraternal twin, as demonstrated in Figures 9.5 and 9.6 for the same hexagons as in Figures 9.3 and 9.4. Again, the alternative dissection comes from replacing each cut of length y in the z-hexagon by a cut from the same vertex but at an angle of 60° from the original cut. Again, dotted lines indicate these cuts in the z-hexagon of Figure 9.3. The triangular pieces in Figure 9.6 are much smaller than any piece in Figure 9.3. However, the more important feature is that the triangles are right triangles, and this makes it easier to produce analogous dissections for any regular polygon with an even number of sides.

If p is an even integer greater than 4, then there are $(p + 1)$-piece dissections of $\{p\}$s for $y/x = 2n \sin(2\pi/p)$, for every whole number n. Although these dissections comprise the main branch of the family, they were not easily detectable

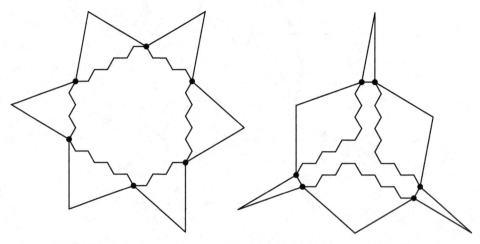

9.4: Hinges for hexagons   **9.5:** Hinges for alternative hexagons

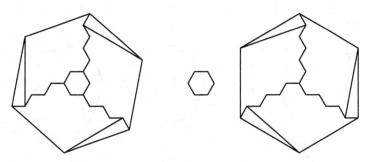

9.6: Alternative hexagons for $1^2 + (\sqrt{27})^2 = (\sqrt{28})^2$

and did not include dissections for odd values of p. To emphasize the stealth that cloaked the members of this much more general class, I call it the *X-gon* class.

Method X-gon: For $\{p\}$s in the X-gon class.

Orient the y-$\{p\}$ so that the top vertex is directly above the center.

Repeat $\lfloor p/2 \rfloor$ times:

    Step from the top vertex:

        Move/cut down-right 1 at an angle of π/p from the vertical.

        Mark the current position.

        Move/cut down-left 1 at an angle of π/p.

        $[n-1$ times$]$:

          $\{$Move/cut down-right 1 at an angle of π/p.

          Move/cut down-left 1 at an angle of π/p.$\}$

    Move/cut from the last mark to the vertex

        that is counterclockwise from the top vertex.

    Rotate the y-$\{p\}$ polygon $4\pi/p$ radians clockwise.

If p is odd, then

    Move/cut from the center to the midpoint of the edge

        to the right of the top vertex.

    Move/cut from the top vertex to a point on the last cut

        that is $x/2$ from the edge that is bisected.

Of course, $2\sin(2\pi/6) = \sqrt{3}$, $2\sin(2\pi/8) = \sqrt{2}$, and $2\sin(2\pi/12) = 1$. Once I had stumbled upon the dissections for $p = 8$ (see Figures 9.7 and 9.8) and $p = 12$ (Figures 17.5 and 17.6) for the case $n = 1$, I recognized the pattern. I describe this new approach in Method X-gon.

Note that the angle between the cuts in the step on either side of the marked position is precisely the angle of the polygon. This fact ensures that angles of the

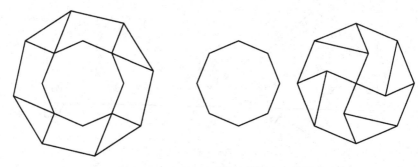

9.7: Octagons for $1^2 + (\sqrt{2})^2 = (\sqrt{3})^2$

z-polygon are formed correctly. It also guarantees that the triangles formed are right triangles. The last five lines of Method X-gon allow us to also handle regular polygons with an odd number of sides. We will look at examples of these in a moment.

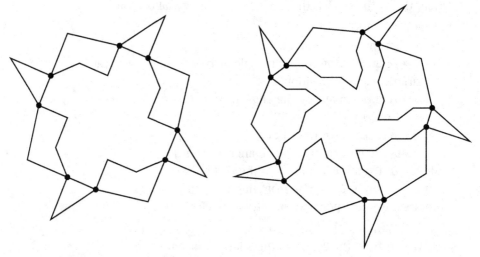

9.8: Hinges for octagons   **9.9:** Hinges for decagons

As a further example, taking $p = 10$ gives decagons. With $n = 2$, we get the hingeable dissection in Figure 9.10. The hinged pieces are in Figure 9.9. All of these dissections are hinge-snug.

Originally, I did not see how to extend Method X-gon to apply to regular polygons with an odd number of sides. However, there is a nice trick that works when p is an odd integer greater than 3. The trick gives $(p + 2)$-piece dissections of $\{p\}$s for $y/x = 2n \sin(2\pi/p)$, for every whole number n. The method handles all but one side in the same way as even polygons – it uses the trick merely on the last side.

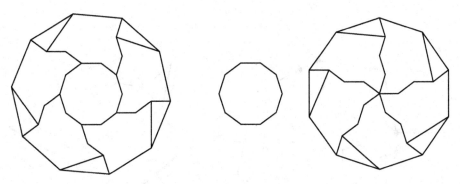

9.10: Decagons for $y/x = 4\sin(\pi/5)$

The trick adapts the technique that the English geometer Harry Hart (1877) used for dissections such as those in Figures 19.1 and 19.2. For a polygon that can be circumscribed by a circle, Hart cut along a perpendicular bisector to the midpoint of each side of the polygon. He then cut along a line from each vertex to the next perpendicular bisector in a counterclockwise direction, cutting off a right triangle whose legs are in the ratio of the side lengths of the polygons that he combined. In Method X-gon, I make $\lfloor p/2 \rfloor$ zigzag cuts from the center of the y-$\{p\}$ to every second vertex as well as $\lfloor p/2 \rfloor$ cuts from a marked vertex to other vertices. Finally, I employ the trick adapted from Hart: For the vertex not yet cut, I cut from the center to the midpoint of one of its adjacent sides and cut off a right triangle pointing in the same direction as the ones made by Method X-gon.

What a lovely group of in-laws we get by marrying Hart's technique to our zigzag! I illustrate this for heptagons with $n = 2$ in Figures 9.11 and 9.12. The dissection of enneagons with $n = 1$ steps forward in Figures 9.14 and 9.13. These dissections are hinge-snug.

Thus, through the match with Harry Hart, we bring together dissections for special cases of every regular polygon except triangles and squares. For $\{p\}$ with

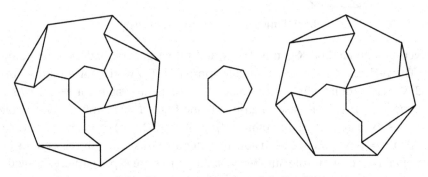

9.11: Heptagons for $y/x = 4\sin(2\pi/7)$

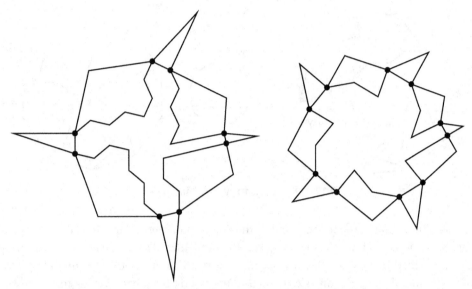

9.12: Hinges for heptagons   **9.13:** Hinges for enneagons

p even, we use $p + 1$ pieces. When p is odd, we use $p + 2$ pieces. So why are triangles and squares penalized? But they are not! Recall Figures 4.3, 4.4, and 5.1. These handle all ratios of y/x and are hingeable. However, neither dissection of squares is cyclicly hingeable.

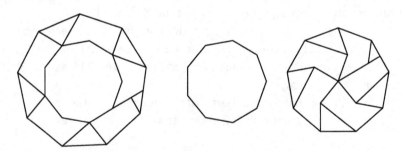

9.14: Enneagons for $y/x = 2\sin(2\pi/9)$

Even though Method X-gon is pretty and rather general, let's not get complacent. The real shocker is that a similar approach also works for all stars $\{p/q\}$. Alfred Varsady found an 11-piece dissection of pentagrams for $(\sin(2\pi/5))^2 + (\cos(2\pi/5))^2 = 1^2$ for the case when $n = 1$, and I noted that similar $(2p + 1)$-piece dissections exist for $\{p/2\}$ for $(\sin(2\pi/p))^2 + (\cos(2\pi/p))^2 = 1^2$, again for the case when $n = 1$. Now I have been able to generalize this group of dissections both for $q > 1$ and for every whole number n, so that there are $(2p + 1)$-piece dissections of $\{p/q\}$ for $y/x = 2n\cos((q-1)\pi/p)\sin(\pi/p)/\cos(q\pi/p)$. I call this the *X-star* class. Varsady's dissection makes a grand entrance in Figure 9.15, accompanied by the hinged pieces in Figure 9.16. If we allow the hinges to be wobbly, then we can cyclicly hinge the pieces. These dissections are also hinge-snug.

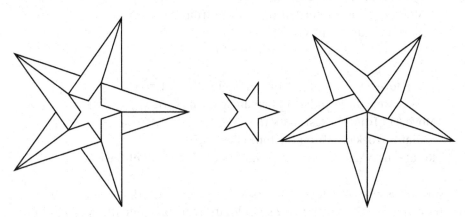

9.15: Varsady's pentagrams for $y/x = \tan(2\pi/5)$

Method X-star describes the approach for stars. For two cases, I can beat the method. For $\{6/2\}$ with $n = 1$, the method gives a 13-piece hinged dissection, but I have already given a 12-piece hinged dissection in Figures 5.16 and 5.17. For $\{8/2\}$ with $n = 1$, the method gives a 17-piece dissection (Solution 17.1), but I give a 16-piece dissection in Figures 17.3 and 17.4. Also, for any pair of hexagrams where $y/x > \sqrt{3}$, there is a 13-piece unhingeable dissection.

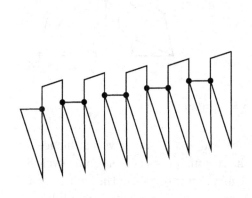

9.16: Hinges for pentagrams

9.17: Hinges for $\{7/2\}$s

Method X-star: For stars $\{p/q\}$ in the X-star class.

Orient the y-$\{p/q\}$ so that a reflex angle is directly above the center.

Repeat p times:

  Step from the top reflex angle:

    Move/cut down-right 1 at $(q-1)\pi/p$ from the vertical.

    Mark the current position.

    Move/cut down-left 1 at an angle of $(q-1)\pi/p$.

    $[n-1$ times]:

      {Move/cut down-right 1 at an angle of $(q-1)\pi/p$.

      Move/cut down-left 1 at an angle of $(q-1)\pi/p$.}

  Move/cut from the last mark to the vertex

    that is clockwise from the top reflex angle.

  Rotate the y-$\{p/q\}$ star $2\pi/p$ radians clockwise about the center.

Now this is a family with connections! Let's look at another example with $q=2$. The dissection for $\{7/2\}$ with $n=2$ sparkles in Figure 9.18, and the cyclically hinged pieces swing along in Figure 9.17. The angle between the cuts in the step on either side of the marked position equals the reflex angle of the star. This ensures that angles in the z-star are formed correctly.

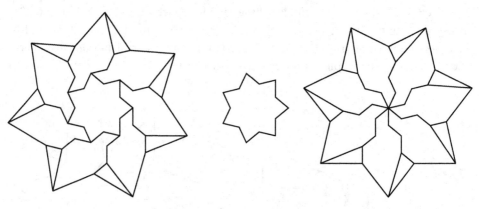

9.18: $\{7/2\}$s for $y/x = 2\tan(2\pi/7)$

Next, let's apply the method to some stars with $q>2$. Along with brilliance in our family comes some instability. All of the resulting dissections can be cyclicly hinged, but we then need wobbly hinges whenever $p < 6(q-1)$. This condition is equivalent to the reflex angle being less than 120°. If we are satisfied with linear hinging, then no wobbly hinges are required. For $\{8/3\}$s where $y/x > \sqrt{2}$, there is a 17-piece unhingeable dissection.

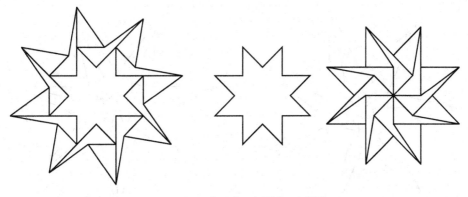

9.19: $\{8/3\}$s for $1^2 + (\sqrt{2})^2 = (\sqrt{3})^2$

9.20: Hinges for $\{8/3\}$s for $1^2 + (\sqrt{2})^2 = (\sqrt{3})^2$

For an $\{8/3\}$ with $n = 1$, my method gives the 17-piece dissection that debuts in Figures 9.19 and 9.20. Since $\sin(\pi/8) = \cos(3\pi/8)$, we have $y/x = \sqrt{2}$, and consequently $z/x = \sqrt{3}$. Another hinged dissection is possible, which makes a nice puzzle.

Puzzle 9.1 Find a 17-piece hingeable dissection of $\{8/3\}$s that is different from the one in Figures 9.19 and 9.20.

For a $\{7/3\}$ with $n = 2$, my method molds the 15-piece dissection of Figures 9.21 and 9.22. What terrific family values! It is nontrivial to verify visually that no wobbly hinges are needed to rotate the hinged pieces from their position in the y-star to their position in the z-star.

There is a remarkable kinship between Method X-star and Method X-gon, at least when the latter is restricted to polygons with an even number of sides. This suggests that the two methods are more related than we might have first believed. And indeed they are! The trick is to recognize the polygon $\{2p\}$ as an appropriately "puffed-out star." If we take $q = 1/2$ then we can still use our notation $\{p/q\}$. Consider a circle with p equally spaced points on it. A $\{2p\}$ is a p-pointed "star" with the points on the circle and sides from each point directed toward a position on the circle halfway between the current point and the next point. The two sides coming from neighboring points meet at the position midway between them on the

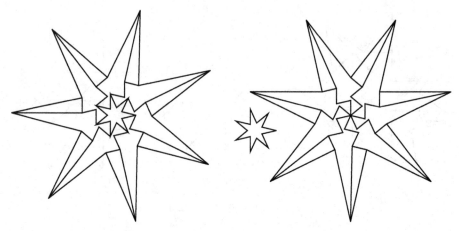

9.21: $\{7/3\}$s for $y/x = 4\cos(2\pi/7)\sin(\pi/7)/\cos(3\pi/7)$

circle, giving a $\{2p\}$. Method X-star on a $\{p/0.5\}$ produces the same dissection as Method X-gon on a $\{2p\}$.

We have now hinted at the generalization of a star to a *pseudostar* $\{p/q\}$, where q need not be an integer but just a real number in the range $1/2 \leq q < p/2$. This generalization is consistent with the definitions of the British industrial chemist Stuart Elliot (1995), who defined and dissected pseudostars that correspond to the case in which q is an odd multiple of $1/2$. In (1999), Robert Banks, a former professor of engineering, generalized pentagrams in a manner equivalent to what we have described here. Banks noted that the nineteenth-century German cartographer Heinrich Berghaus projected the world onto a pentagram whose angles are approximately 52.53°. With some calculation, we see that this translates to a $\{5/q$

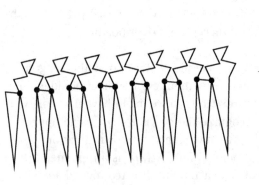

9.22: Hinges for $\{7/3\}$s

9.23: Hinges for $\{5/1.7\}$s

with $q \approx 1.77$. The first such star projection was a modification by August Peter-mann of an 8-pointed star by Dr. G. Jäger (1867), the director of the Vienna zoo. In these pseudostars, the distance from the center to a point is twice the distance from the center to a reflex angle. It follows that $\cos((q-1)\pi/p) = 2\cos(q\pi/p)$. For the Jäger–Petermann 8-pointed star, $q \approx 3.13$.

Can we bring pseudostars into our extended family? Method X-star works just fine on a pseudostar. We see it applied to a $\{5/1.7\}$ with $n = 1$ in Figure 9.24; the hinged pieces are in Figure 9.23.

9.24: $\{5/1.7\}$s for $y/x = 2\cos(0.7\pi/5)\sin(\pi/5)/\cos(1.7\pi/5)$

What happens if we apply Method X-star to pseudostar $\{p/1\}$? Then we obtain an instance of Harry Hart's dissection (Figure 19.1) applied to regular polygons $\{p\}$. Since $q - 1 = 0$ in this case, the zigzag cuts disappear and are replaced by straight cuts. Then n need no longer be a whole number, and the polygons need no longer be regular, as we shall see in Chapter 19. What a wonderful family portrait!

Boy, the way Dudeney played.
  Challenges that did not fade.
Guys like us were not dismayed.
  Those were the days.

And we cut up more than once.
  'Grams to 'grams and 'gons to 'gons.
Mister, we could use a man
  like Harry Lindgren at once.

Didn't need no theorem state'.
  Every figure 'lucidate.
Gee, our hinges swung so great.
  Those were the days.

Curious Case, part 3: Trouble with Attributions

Henry Dudeney was not always careful in attributing puzzles and their solutions to their original sources. In this regard he may have been influenced by the puzzle "literature" of his time, in which wholesale borrowing from other sources – without any attribution at all – seemed to be the norm. His approach probably also reflected the fact that he was writing a puzzle column for a weekly newspaper and not an article for a scholarly journal. We shall review examples that lend credence to concerns about whether Dudeney discovered the triangle-to-square dissection himself.

I will give examples from his dissection problems primarily because I am more familiar with the history of these puzzles. The following four examples appeared in the *Weekly Dispatch* between 1896 and 1903.

1. In "The Greek Cross Puzzle" on August 2, 1896, Dudeney gave a 4-piece dissection of a Greek Cross to a square. This was the same 4-piece dissection that Don Lemon (1890) had given. There is no attribution or explanation in Dudeney's column.

2. In "The Three Squares Puzzle" on April 8, 1900, Dudeney noted that various puzzle books had given a 7-piece dissection for dissecting three equal squares to one, whereas he gave a solution in six pieces. However, Henry Perigal (1891) had already given the same 6-piece dissection. Moreover, Perigal's dissection is remarkably similar to the 4-piece dissection of a gnomon to a square by Philip Kelland (1855). The gnomon consists of three equal squares. If we cut Kelland's dissection along the boundaries of these squares, we obtain precisely Perigal's dissection. Again there is no attribution in Dudeney's column.

3. In "The Arbour Table" on July 28, 1901, Dudeney gave a 5-piece dissection of a hexagon to a square. This is the same dissection that Édouard Lucas (1883) attributed to Paul Busschop, who had claimed it in (1876). Dudeney was aware of Busschop's work, at least by 1903 when he mentioned him in another puzzle. Dudeney was also familiar with Lucas's work. Again, there is no attribution by Dudeney.

4. In the solution to "The Pentagon and Square" on March 15, 1903, Dudeney gave a 6-piece dissection of a regular pentagon to a square. He claimed that, up to that time, the best solution had been in 7 pieces. But Robert Brodie (1891) had already given a 6-piece solution. Dudeney noted that Paul Busschop had found the 7-piece solution but then failed to state that the first step in his own dissection – of converting the pentagon to a parallelogram – was due to Busschop. Furthermore, Dudeney connected the solution with what he called "our hexagon puzzle, 'The Arbour Table'," but he failed to state that this solution is the same one that Busschop gave.

A more serious problem with Dudeney's puzzle of the "Arbour Table" arose on July 28, 1901, when he incorrectly explained the solution:

This is undoubtedly one of our harder puzzles, and I am not surprised that it has defeated nearly all of our cleverest competitors. Indeed, I will admit that without some elementary knowledge of geometry the task of finding an exact answer is practically a hopeless one. As it is, the only two competitors who met with any success ....

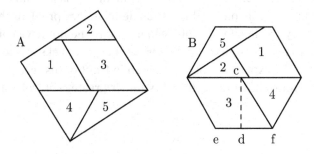

If you cut your square into the five pieces by lines in Diagram A they will fit together as in Diagram B and so form a regular hexagon. The reader will soon see that if he were given both the square and the length of the side of the required hexagon, the problem would be easy enough, for once you have marked off the piece 4 and the piece 3, which is just twice as large, the work is done. But the difficulty is to find the hexagon's side.

Dudeney acknowledged the problem in his column of August 4:

Some of my remarks last week respecting the "Arbour Table" solution were misleading – in fact, obviously incorrect – but I will make the whole matter clear in the next issue.

Finally, on August 11, he wrote:

As stated in our last issue, I found that some of my statements respecting the solution of this puzzle were very misleading. Though the solution was quite correct in intention, some of my explanatory remarks were certainly of a nature calculated to bewilder the geometrically-inclined reader. I wrote the article under the combined influences of heat, haste, and holiday from some old notes of mine ... the piece 4 is not an equilateral triangle, as I said, nor is piece 3 composed of two such triangles....

My misleading description has shown me, by the number of letters I have received on the subject, how carefully readers examine our solutions and take nothing for granted. That is wholly satisfactory and pleasant to know. But, curiously enough, not one of my correspondents succeeded in discovering the right interpretation of the solution. They found that my remarks were wrong, but they did not see how to correct the slip.

We can draw several inferences from this episode. First, Dudeney was evidently not so familiar with this dissection, because he made a rather simple blunder. It appears likely that he had borrowed the dissection from another source and thus was just moderately skillful with dissection problems. Second, only two readers found a minimal solution, and not one of the many readers who detected the flaw in the explanation were able to correct it. Thus he might have realized that dissection problems such as this one and the triangle-to-square problem that he would give later were very challenging for his readers. Could he not have concluded that putting out the triangle-to-square puzzle for another two weeks was in vain? Third, Dudeney had had the experience of slipping up in front of his readers. This might have given him an incentive to keep things under greater control should another slip-up occur.

CHAPTER 10

Elation in Tessellations

Let's return once again to tessellations, using them in this chapter to dissect one figure into another. This time we'll generate some real excitement, maybe even elation,
in tessellations – so much so, that we will soon find ourselves leaping off of the plane
and onto the surface of a cube. Before we get too high, let's recall some ground rules
from Chapter 3. Each symmetry point is a point of rotational symmetry with respect
to each tessellation. When we superpose tessellations, we require a symmetry point
of one tessellation to coincide with a symmetry point of the other tessellation.

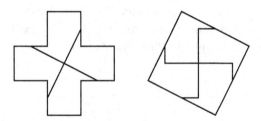

10.1: Loyd's Greek Cross to a square

First, let's dissect a Greek Cross to a square. Sam Loyd (*Tit-Bits,* 1897a) gave
a pretty 4-piece hingeable dissection, shown in Figure 10.1. He cut the cross into
four identical pieces, so that there is 4-fold rotational symmetry in both the cross
and the square. We can derive this dissection by superposing the tessellations in
Figure 10.2; the hinged pieces swing around in Figure 10.3. The dissection is hinge-
snug and grain-preserving.

There is a second hingeable 4-piece dissection of a Greek Cross to square
(Figure 10.5). In (1890), Don Lemon (pseudonym of London banker Eli Lemon Shel-
don) was the first to publish what was essentially this dissection. He cut the Greek
Cross in the same way but formed the square from these pieces in a way that is not
hingeable. The square as shown is consistent with any one of four hingings, one of
which we see in Figure 10.6. The hinging shown is hinge-snug.

We can derive the dissection by using a hexahedral tessellation. The tessellation element is in Figure 10.4. Dashed edges of three different types indicate the

93

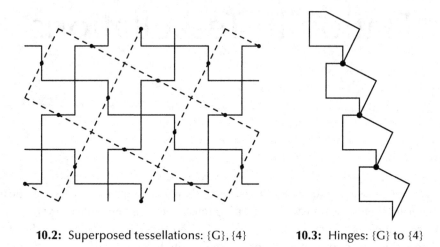

10.2: Superposed tessellations: {G}, {4}   **10.3:** Hinges: {G} to {4}

outline of a square. The edges must be matched in a way that respects the length of dashes and the direction of arrows. There are four types of symmetry points, labeled A, B, C, and D. Point A is a center of 3-fold rotational symmetry, and points B and C centers of 2-fold rotational symmetry. Point D is a center of 1-fold rotational symmetry in the tessellation of crosses and 3-fold symmetry in the tessellation of squares. In a moment, I will justify these claims about points A and D.

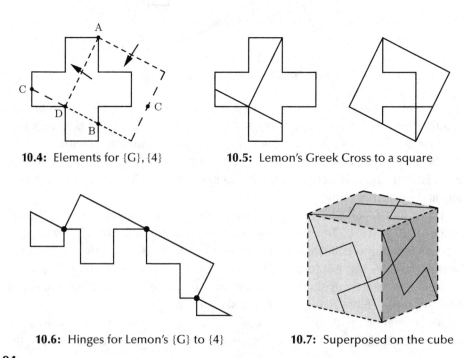

10.4: Elements for {G}, {4}   **10.5:** Lemon's Greek Cross to a square

10.6: Hinges for Lemon's {G} to {4}   **10.7:** Superposed on the cube

Although this tessellation element does not tile the plane, it does tile a cube. Figure 10.8 shows the complete hexahedral tessellation, folded out flat. Arrows indicate how several of the edges fit together to form the cube. Again, the symmetry points are at points of rotational symmetry: one in the middle of the edge represented by short dashes, one in the middle of the edge represented by long dashes, and one at the vertex labeled by A in Figure 10.4. We see the Greek Cross tessellation superposed on the cube in Figure 10.7. Points A and D are at the corners of the cube. The angles of faces that meet at these corners total 270°. Thus, we divide by 270° rather than 360° when determining the n-fold symmetry of points at the corners.

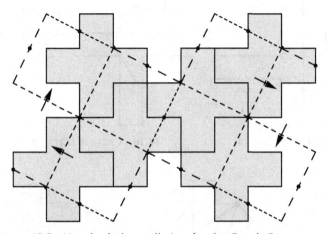

10.8: Hexahedral tessellation for the Greek Cross

So far we have seen two different 4-piece hingeable dissections of a Greek Cross to a square. The first has 4-fold rotational symmetry and the second has no rotational symmetry. There is also a third, which has 2-fold rotational symmetry.

Puzzle 10.1 Find a superposition of the tessellations of squares and of Greek Crosses that leads to a 4-piece hingeable dissection of a Greek Cross to a square, and identify the hinged pieces.

We next consider the Cross of Lorraine, which is created by attaching thirteen equal squares together. Bernard Lemaire proposed dissecting this cross to a square and found an 8-piece unhingeable dissection, as reported by math puzzle columnist Pierre Berloquin (*Monde,* 1974a). Later, Berloquin (*Monde,* 1974c) reported that he had received a 7-piece unhingeable dissection from Mr. Szeps, and he challenged his readers to find it. Berloquin (*Monde,* 1974d) reported that Bernard Lemaire, a professor of operations research in Paris, had then also found a 7-piece unhinge-able dissection, and he gave Szeps's and Lemaire's dissections. Szeps's dissection can be derived by superposing tessellations on a cube.

Now I have found a 7-piece hingeable dissection (Figure 10.10) that is related to Szeps's approach. We can see from the hinged pieces are in Figure 10.11 that the dissection is wobbly hinged. I found the dissection by trial and error but then real-ized that it is derivable by superposing tessellations on the cube (Figure 10.12).

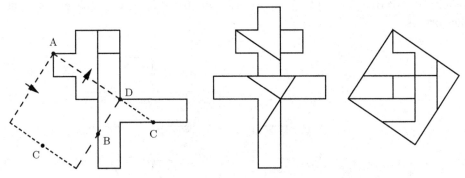

10.9: Elements for {L'}, {4}   **10.10:** Cross of Lorraine to a square

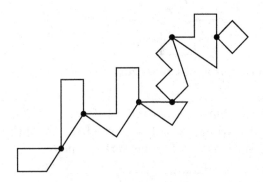

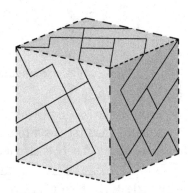

10.11: Hinges for a {L'} to a {4}   **10.12:** Superposed on the cube

The tessellation elements overlap in Figure 10.9. It is easy to see how to rotate the three pieces from their position in the cross. The dashed edges and symmetry points – labeled A, B, C, and D – duplicate precisely their roles in the similar Fig-ure 10.4. We see the complete hexahedral tessellation folded out flat in Figure 10.13. The representation is analogous to that of Figure 10.8. When we examine the super-position in Figure 10.13, we see that a pair of lines cross at one point that is not a

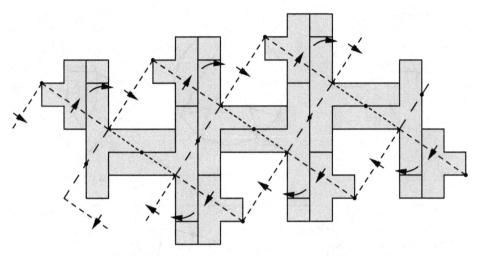

10.13: Hexahedral tessellation for the Cross of Lorraine

symmetry point in either tessellation. As I discussed in Chapter 3, one such bad crossing does not necessarily render the resulting dissection unhingeable, as we see in this case.

Lemaire extended Szeps's dissection to a similar unhingeable dissection of the Cross of Lorraine to a Greek Cross in just 7 pieces. I have extended my dissection to a hingeable dissection of the Cross of Lorraine to a Greek Cross, also in just 7 pieces. This dissection (Figure 10.15) is derivable using hexahedral tessellations. The tessellation elements for the crosses and the square overlay each other in Figure 10.14.

Harry Lindgren (1951) gave a beautiful 6-piece unhingeable dissection of a dodecagon to a square. My 8-piece hingeable dissection is related to Lindgren's

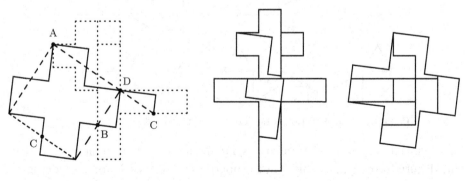

10.14: Elements for {L'}, {G}   **10.15:** Cross of Lorraine to a Greek Cross

97

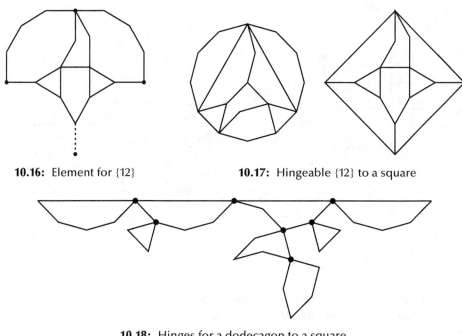

10.16: Element for {12}   **10.17:** Hingeable {12} to a square

10.18: Hinges for a dodecagon to a square

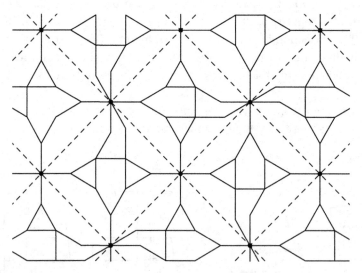

10.19: Superposition of tessellations for a dodecagon to a square

unhingeable dissection. As in Lindgren's, I found a tessellation element. Of course, mine (Figure 10.16) is hingeable. Then I superposed the tessellation of dodecagons with a tessellation of squares (Figure 10.19), giving the 8-piece hinged dissection in Figures 10.17 and 10.18.

Small dots indicate the points of rotational symmetry in the tessellations. Half of these points have 2-fold rotational symmetry, and the remaining ones have 4-fold symmetry. We could hinge the same pieces in other ways. For example, we could connect the lowest piece by attaching its upper right vertex to the lowest vertex of the triangle on the right. Also, we could connect the four large pieces by three hinges in several different ways. All possible hingings seem to have a pair of abutting hinges.

Puzzle 10.2   Find an 8-piece hingeable dissection of a Greek Cross to a dodecagon.

To bring this chapter to a euphoric conclusion, let's try a dissection of a {12/2} to a hexagon. Lindgren (1964b) gave a 10-piece unhingeable dissection, and in (Frederickson 1972d) I gave an 8-piece unhingeable dissection. Both of these dissections derive from tessellations in which the superposition splits one large piece into four pieces that we can hinge together.

To find a completely hingeable dissection, I first designed the hingeable tessellation element in Figure 10.20. This element imposes an internal structure on the

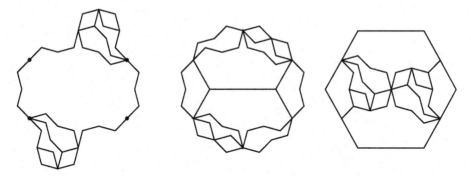

10.20: Element for {12/2}   **10.21:** Hingeable {12/2} to a hexagon

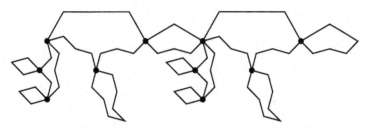

10.22: Hinges for a cross of a {12/2} to a hexagon

99

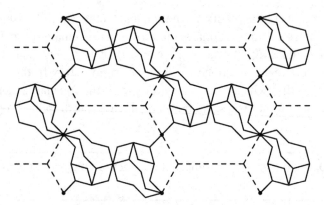

10.23: Superposition of tessellations for a {12/2} to a hexagon

{12/2} that is richer than one consisting only of rhombuses. The superposition in Figure 10.23 produces the 12-piece hingeable dissection that graces Figure 10.21. The hinged pieces in Figure 10.22 seem almost to float, as if symbolic of our tessellation technique, which has reached new heights.

CHAPTER 11

Strips Revealed

In the first few chapters, we have seen several hingeable dissections produced using the strip technique. Now it is time to search with enthusiasm for such dissections, focusing on those of one figure to another. Hopefully we can lay bare a variety of ways to coax these shy dissections into the light of day. First, we will review a few early dissections that turn out to be hingeable, although their originators may not have realized all of their secrets. Then we will strike out into virgin territory and discover a number of captivating dissections. To facilitate crosspositions and avoid cross puzzlers, I promise to limit the further use of questionable metaphors.

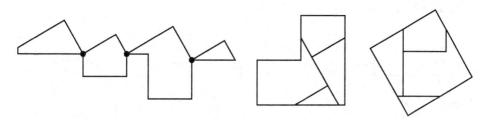

11.1: Hinges: gnomon to square **11.2:** Kelland's gnomon to a square

Philip Kelland (1855, 1864) gave many 4-piece dissections of a square to a gnomon created by attaching three squares together. Interestingly, one of these is hingeable, number XXII in (1864). The hinged pieces swing open in Figure 11.1, next to the dissection in Figure 11.2. The TT2 crossposition in Figure 11.3 generates it. This dissection is hinge-snug and grain-preserving.

Puzzle 11.1 Find a different 4-piece hingeable dissection of a gnomon to a square that uses a TT2 crossposition.

A *mitre* results when we remove one of the four isosceles right triangles from the large square in Figure 1.2. Sam Loyd (*Inquirer,* 1901a) published an erroneous 4-piece dissection of a mitre to a square. Taking great delight in pointing out Loyd's

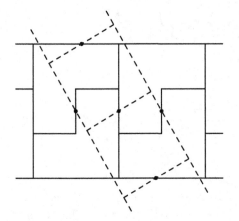

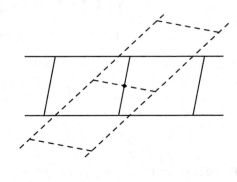

11.3: Crossposing gnomons, {4}s

11.4: Crossposing parallelograms

misstep, Henry Dudeney (*Strand*, 1911a) gave a correct 5-piece dissection. Dudeney's dissection is unhingeable, but we can adapt Kelland's dissection of the gnomon to a square to give a 5-piece hingeable dissection. Just cut a diagonal from the lower left corner of the gnomon to its reflex angle and then rotate the resulting piece flush against the other edge of the reflex angle, giving a mitre. We can apply the same adaptation to Figure 3.19.

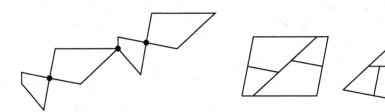

11.5: Hinged pieces

11.6: Taylor's parallelogram to parallelogram

Henry Taylor (1905) gave a hingeable 4-piece dissection of a parallelogram to another parallelogram, an example of which is shown in Figure 11.6. The dissection is rather attractive, as each parallelogram displays 2-fold rotational symmetry. Taylor did not directly state that his dissection was hingeable. However, he did specify that two of the pieces should be rotated through two right angles around points that we identify as hinge points. The hinged pieces are shown in Figure 11.5. The TT1 crossposition that leads to this dissection precedes it in Figure 11.4. This dissection is also hinge-snug and grain-preserving.

The next dissection in Taylor's paper is a 3-piece dissection of a parallelogram to another parallelogram. This latter dissection corresponds to a plain-strip dissection. Since Taylor was acquainted with the more economical dissection, he may have included the 4-piece dissection because he recognized its additional properties. There are other ways to produce a hingeable dissection of one parallelogram to another, as readers can infer from Figure 11.17.

William Macaulay (1915) investigated the conditions under which we can produce 8-piece dissections of one quadrilateral to another. He identified a nifty dissection that we can derive by using strips. First we cut one quadrilateral as in Figure 11.7, where the endpoints of the cuts are midpoints of the sides. If we swing around the triangles, we will get a P-strip element. The resulting strip is shown in solid in Figure 11.8, where it crossposes with another strip derived from the other quadrilateral. The key to the dissection is to crosspose the two strips so that a point from one strip in which several vertices meet coincides with a similar point from the other strip.

The resulting dissection in Figure 11.9 is irregular and yet still lovely, as we can see from its hinged pieces in Figure 11.10. This PP dissection is somewhat unique in

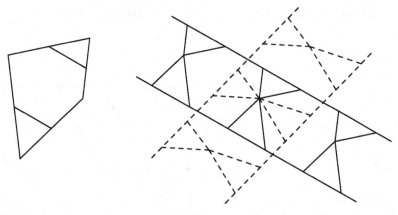

11.7: Cutting the quad   **11.8:** Crossposition of quadrilaterals

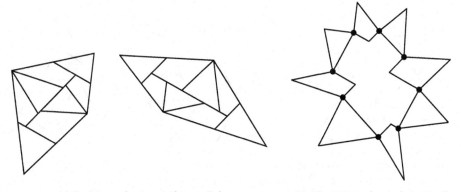

11.9: Macaulay's quad to quad   **11.10:** Hinged quad to quad

103

that it uses no anchor points as symmetry points. Instead, it relies on the fact that the triangles created by cuts as in Figure 11.7 can rotate either clockwise or counterclockwise. Indeed, we use both endpoints of the cut, which allows us to hinge the pieces cyclicly. Thus we get away with line segments that meet at a point that is interior to both strips but is not an anchor point. This dissection is hinge-snug and grain-preserving.

Aside from the square, the hexagon is arguably the regular polygon most conducive to strip dissections. Paul Busschop (1876) discovered an unhingeable 5-piece dissection of a hexagon to a square. We can derive his dissection by slicing the hexagon in equal halves along a diagonal, forming a P-strip from the two pieces, and then crossposing that strip with a strip of squares. We can then base a hingeable T-strip element on the same two halves of the hexagon, as in Figure 11.11, which precedes the TT-strip crossposition in Figure 11.12.

My corresponding 6-piece hingeable dissection of a hexagon to a square is in Figure 11.14; the hinged pieces appear in Figure 11.13. This dissection is hinge-snug. The boundary of the strip of squares crosses a line segment that is on the interior

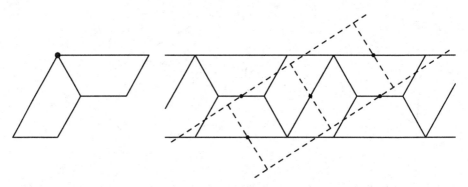

11.11: {6} element   **11.12:** Crossposition of hexagons and squares

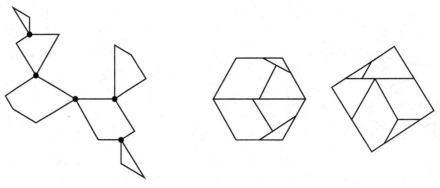

11.13: Hinges for {6} to {4}   **11.14:** Hexagon to a square

of the hexagon strip. Thus we have another exception to requiring that all intersections be at symmetry points.

There are a number of other possible 6-piece hingeable dissections, which leads to the question of how we might choose among them. One criterion is whether a dissection avoids small pieces. Another is whether it is symmetric. Roman architect Duilio Carpitella (1996) gave two 6-piece hingeable dissections, although it is not clear whether he was aware that either is hingeable. His dissections have 2-fold rotational symmetry.

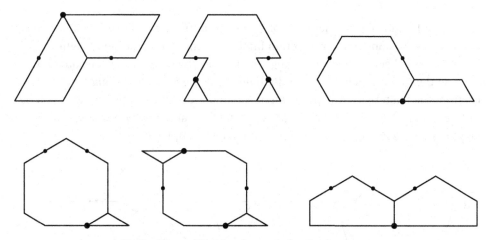

11.15: Hinged T-strip elements for the hexagon

In Figure 11.15, I give six hingeable T-strip elements for hexagons. In each element, the large dots represent points where the pieces are hinged. The small dots indicate anchor points for the T-strip. We have already seen the element in the upper left in Figure 11.11. We can use the element in the lower middle to derive Carpitella's dissections. Did you see that the element in the upper middle is wobbly hinged?

We now move on to hingeable dissections of a pentagon to a square. Busschop (1876) found a 7-piece unhingeable dissection, and Scottish engineer Robert Brodie (1891) found a 6-piece unhingeable dissection. To derive a hinged pentagon P-strip element, we slice an isosceles triangle off the top of the pentagon, leaving a trapezoid. Next, we split the triangle into two triangles, swinging one to the right and the other to the left; the one on the right combines with the trapezoid to give a parallelogram. Then we cut the triangle on the left into three pieces that rearrange to give a narrow parallelogram. All of the angles created are multiples of 36°. The hinged pieces in the P-strip element open up in Figure 11.16, and a PT crossposition with a strip of squares follows in Figure 11.17. This gives an 8-piece hingeable dissection that is hinge-snug. But we can do better!

105

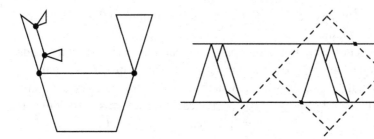

11.16: {5} unfolded   **11.17:** Crossposition of pentagons and squares I

We will find a T-strip element for a pentagon that we can use in a TT2 cross-position. Let's examine the dissection implied by the crossposition in Figure 11.17. There is a small acute isosceles triangle just below the top piece of the square. Can we glue these two together? If we do, then the acute isosceles triangle now sticks up from the main piece; the partition of the upper portion of the pentagon is a bit different so as to produce this isosceles triangle in an appropriate position, as in Figure 11.18. Again, all of the angles created are multiples of 36°.

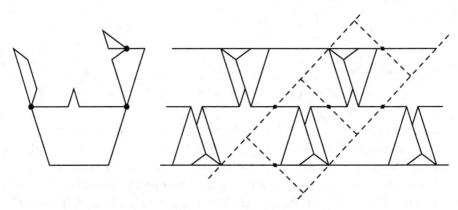

11.18: {5} unfolded II   **11.19:** Crossposition of pentagons and squares II

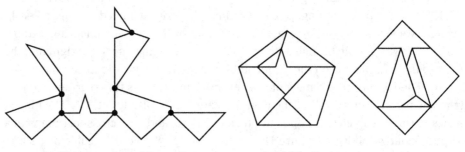

11.20: Hinges for {5} to {4}   **11.21:** Pentagon to a square

The crossposition in Figure 11.19 just barely avoids a bad crossing. There is an edge that is approximately 0.1% of the pentagon's side length. But even if there had been a bad crossing, we could have crossposed the square strip the other way. This near coincidence of two points suggests that there may be only a limited use of the strip: to squares. I show the 7-piece hingeable dissection in Figure 11.21, with one of four possible hingings in Figure 11.20.

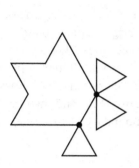

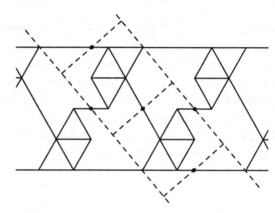

11.22: {6/2} unfolded   **11.23:** Crossposition: hexagrams and squares

Dissecting a hexagram to a square is a problem that attracted attention relatively early. An 8-piece unhingeable dissection appeared in the anonymous Persian manuscript, *Interlocks of Similar or Complementary Figures,* from approximately 1300. Sam Loyd (*Eliz. J.,* 1908a) gave an unhingeable 7-piece dissection of a hexagram to a square, and Harry Bradley (1921) found a 5-piece unhingeable dissection. I have found the 4-piece T-strip element for a hexagram in Figure 11.22. The TT2 crossposition with the strip of squares in Figure 11.23 gives the 7-piece hingeable dissection in Figure 11.24. The hinged element, and thus the hinged pieces in Figure 11.25, use a triple hinge. This dissection is wobbly hinged.

Let's consider some dissections that don't involve squares, such as a hexagon to a triangle. A 9-piece unhingeable dissection appeared in the circa 1300 *Interlocks*

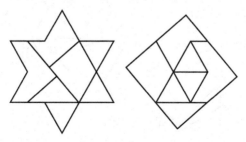

11.24: Hingeable dissection of a hexagram to a square

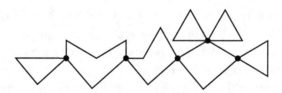

11.25: Hinges for a hexagram to a square

manuscript. In (1940), Michael Goldberg, a mathematician who worked for the U.S. Navy, gave a 6-piece unhingeable dissection, and Harry Lindgren (1951) found a 5-piece unhingeable dissection. Carpitella (1996) gave a 6-piece hingeable dissection, but it seems that he was not aware that it is hingeable. I give a different 6-piece hingeable dissection, using the T-strip for hexagons from Figure 11.11. The cross-position is shown in Figure 11.26, the hinged pieces in Figure 11.27, and the hinge-snug dissection in Figure 11.28. Do you see where the boundary of the hexagon strip crosses an interior edge of the triangle strip?

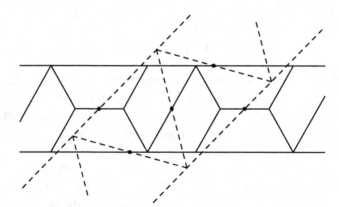

11.26: Crossposition of hexagons and triangles

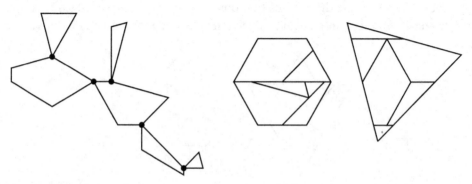

11.27: Hinges for {6} to {3}

11.28: Hexagon to a triangle

How about dissecting a pentagon to a triangle? Goldberg (1952) gave a 6-piece unhingeable dissection. For a hingeable dissection, we could use the P-strip element from Figure 11.16, but that element has smaller pieces than my new element in Figure 11.29. I convert the top triangle of the pentagon into a parallelogram with two cuts. Once again, all of the angles created are multiples of 36°. Then I cut an isosceles triangle out of the large piece to make room for the parallelogram. The PT crossposition in Figure 11.30 produces the 8-piece hinge-snug dissection in Figures 11.31 and 11.32.

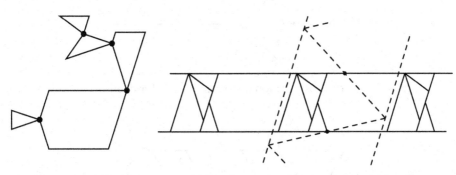

11.29: {5} unfolded III   **11.30:** Crossposed pentagons and triangles

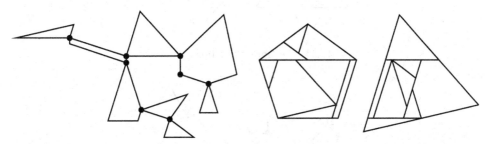

11.31: Hinges for {5} to {3}   **11.32:** Pentagon to a triangle

Having found hinged dissections involving hexagons and pentagons, it is natural to ask if there is a nice hinged dissection from a hexagon to a pentagon. I have found a 10-piece dissection, which I leave as a puzzle.

***

Puzzle 11.2   Find a hinged dissection of a hexagon to a pentagon.

***

The dissection of a hexagram to a hexagon proved a nifty challenge. Bruce Gilson (in Gardner 1969) and independently Lindgren (1964b) found 7-piece unhingeable dissections. My 6-piece unhingeable dissection, with two pieces turned over, first appeared in the second edition of Gardner (1969). I have found a 9-piece hingeable

109

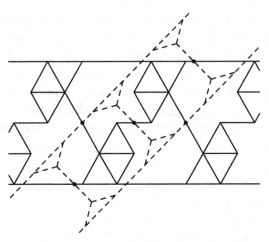

11.33: Crossposition of hexagrams and hexagons

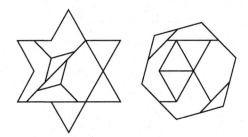

11.34: Hexagram to a hexagon

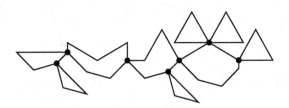

11.35: Hinges for a hexagram to a hexagon

dissection, using the previous hexagram T-strip element and a hexagon element from Figure 11.15. The crossposition in Figure 11.33 yields my hingeable dissection in Figure 11.34. The dissection is wobbly hinged, and also uses a triple hinge, as shown in Figure 11.35. There are many different hingings of these pieces.

Having achieved some measure of success, let's get daring and try an octagon to a hexagon. Lindgren (1964b) gave a 9-piece unhingeable dissection. Gavin Theobald, a British computer programmer, found an 8-piece unhingeable dissection. I have found a hingeable dissection based on an octagon strip derived in the following way. Orient the octagon so that a longest diagonal is horizontal. Imagine slicing half of a 45°-rhombus off the bottom and another half off the top. We cut the two half-rhombuses into pieces that we then place against the sides.

In order to obtain a hingeable construction, we cut the bottom rhombus into four pieces. We need cut the top into only two pieces if we leave a peak in the octagon and a corresponding valley at the top of the sides. We then mate the peaks

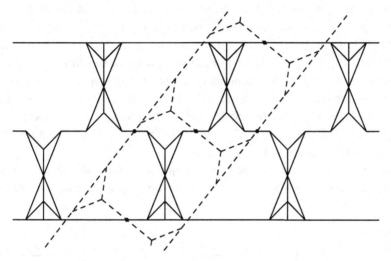

11.36: Crossposition of octagons and hexagons

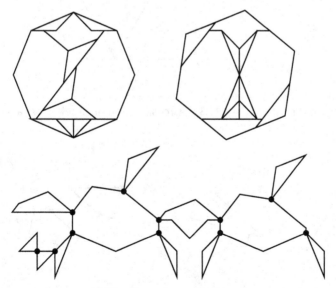

11.37: Octagon to a hexagon

111

and valleys to get a T2-strip. The crossposition of this octagon strip with a hexagon strip in Figure 11.36 leads to the singular hinge-snug dissection in Figure 11.37. The hinged pieces look rather like a "mating dance of two crabs"!

Let's move on to dissections of Greek Crosses. Six months after the triangle-to-square dissection, Henry Dudeney (*Dispatch,* 1902a) discovered how to modify it to give a 6-piece dissection of the Greek Cross to a triangle. Although he apparently did not use T-strips, we can use them to produce his dissection. We slice a Greek Cross into two pieces that form the T-strip element, as shown in Figure 11.38. We then crosspose the resulting T-strip with a strip of triangles, as shown in Figure 11.39. This gives the 6-piece TT2 dissection of Figure 11.40, which we can hinge as in Figure 11.41. The piece in the upper right corner of this latter figure appears to be two separate pieces, but it is actually one piece with a very narrow isthmus, as we can see in the enlargement on the right. Four of the pieces are cyclicly hinged, and the dissection is grain-preserving. Again, the crossposition is exceptional in that edges in the interior of both figures cross. There is a 5-piece unhingeable dissection created by Lindgren (1961).

Before we leave Greek Crosses, let's enjoy a remarkable interaction with a hexagon. Lindgren (1964b) gave several 7-piece unhingeable dissections of a Greek

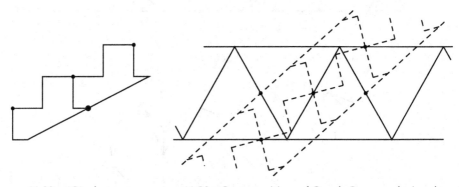

11.38: {G} element   **11.39:** Crossposition of Greek Cross and triangle

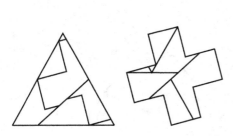

11.40: Dudeney's Greek Cross to {3}

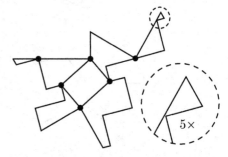

11.41: Hinges for {G} to {3}

Cross to a hexagon. I have found an 8-piece hingeable dissection by using the hexagon element in the lower right corner of Figure 11.15, along with the preceding Greek Cross element, to produce the TT1 crossposition in Figure 11.42. The hinged dissection glistens in Figures 11.43 and 11.44. How lovely that the cuts in both the Greek Cross and the hexagon have 2-fold rotational symmetry. The dissection is grain-preserving, and the pieces are cyclicly hinged. It is not so easy to determine by inspection that there are no wobbly hinges. In fact, I resorted to experiments with a hinged model to help me resolve the issue. To swing open the hexagon, keep the three pieces on the lower right of the hexagon together, and similarly for the three pieces on the upper left. This opens a hole in the shape of a parallelogram.

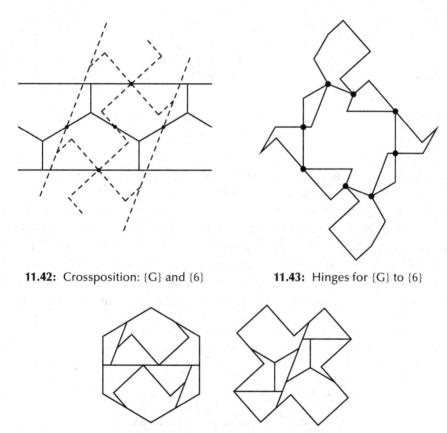

11.42: Crossposition: {G} and {6}

11.43: Hinges for {G} to {6}

11.44: Greek Cross to a hexagon

Let's now progress from Greek to Latin Crosses. Lindgren (1964b) gave several 5-piece unhingeable dissections of a Latin Cross to a square. I have found a 7-piece hingeable dissection. To make a strip element, I cut two small squares and swing them around as shown in Figure 11.45. The PT crossposition of strips in

113

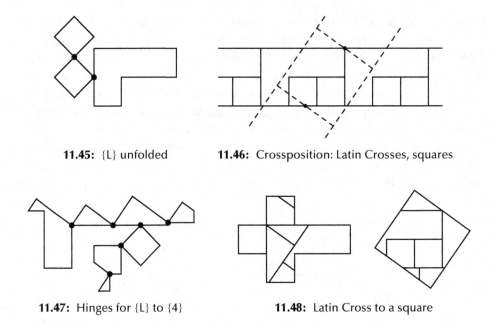

11.45: {L} unfolded   **11.46:** Crossposition: Latin Crosses, squares

11.47: Hinges for {L} to {4}   **11.48:** Latin Cross to a square

Figure 11.46 seems to work best, producing the hinged dissection in Figures 11.47 and 11.48. Both this dissection and the next one have exceptional crossings.

We next find a dissection of a Latin Cross to a triangle. Lindgren (1964b) gave a 5-piece unhingeable dissection, which is similar to his (1961) 5-piece unhingeable dissection of a Greek Cross to a triangle. In both dissections, Lindgren cut one of the arms off of the cross to make a T-strip that was 2 units wide. The latter dissection had beaten Dudeney (*Dispatch,* 1902a) by a piece. But if we look for an analogue of Dudeney's dissection, we are disappointed, since no similar T-strip element seems to be forthcoming. Instead we use the same basic idea as in

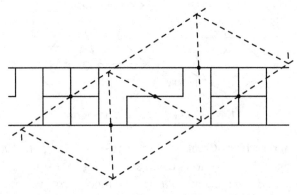

11.49: Crossposition of Latin Crosses and triangles

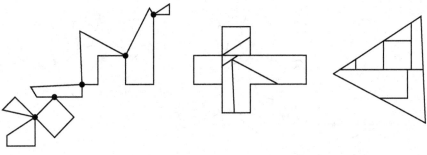

11.50: Hinges for {L} to {3}   **11.51:** Latin Cross to a triangle

Figure 11.45 but swing the second small square differently to form a T-strip element, as we can see in Figure 11.49. A hinged 7-piece dissection is the result in Figures 11.50 and 11.51. Interestingly, the dissection has a triple hinge even though the strip element does not.

As a final treat, let's see what we can do with a {12/2}. Anton Hanegraaf found a 10-piece unhingeable dissection of a {12/2} to a square. Robert Reid (1987) found a bumpy plain strip that gave a 9-piece unhingeable dissection with one piece turned over, and Gavin Theobald improved that to eight pieces. I have found a 10-piece

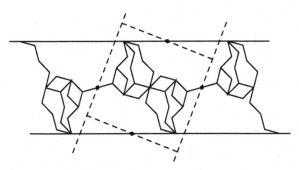

11.52: Hingeable crossposition of {12/2}s and squares

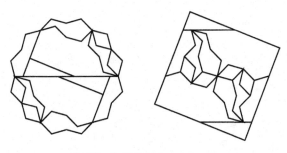

11.53: A {12/2} to a square

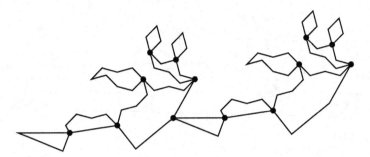

11.54: Hinges for a {12/2} to a square

T-strip element for a {12/2}, as in the TT2 crossposition with a strip of squares in Figure 11.52. I derive the T-strip from the tessellation of a {12/2} that appears in Figure 10.23. The crossposition gives the 14-piece hinged dissection in Figures 11.53 and 11.54.

Turnabout 3: Hinged Tessellations

A *hinged tessellation* is a tessellation to which we add hinges in a regular fashion. We can then pull open the tessellation, rotating some of the figures relative to others, so that the background behind the tessellation's figures becomes visible. Of course, we cannot construct a complete model of a hinged tessellation, because we cannot assemble an infinite number of figures. Duncan Stuart (1963) and Ron Resch (1965) discovered these independently and gave several examples. David Wells (1988, 1991) duplicated two of the examples and called them hinged tessellations. As beautiful as they are, hinged tessellations do not seem to be of any use in creating hinged dissections. However, Helena Verrill (1998) discussed the use of hinged tessellations for finding origami tessellations.

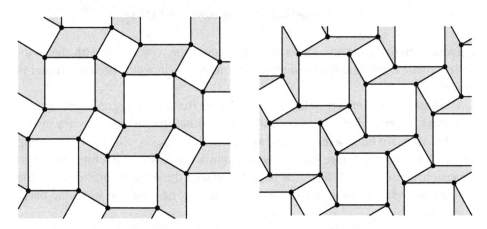

T8: Hinged tessellation of a pair of unequal squares

We explore three representative examples of hinged tessellations here. Both Duncan Stuart and Ron Resch gave the tessellation of equal squares as an example of a hinged tessellation. It is natural to generalize this to the tessellation of a pair of unequal squares in Figure T8. Here the squares are in white, and the shaded parallelograms indicate the holes between the squares. I show two orientations of the hinged tessellation. As the small squares rotate relative to the large ones, the tessellation can change from one in which the parallelograms have angles of 0° and 180° to one in which the parallelograms are rectangles and finally to one with trivial parallelograms once again. Winkler (1929) derived a hinged mechanism that leads in a natural way to this hinged tessellation.

We can also tessellate the plane using a hexagram and a triangle. When we hinge every second point of the hexagram with the corner of a triangle, we obtain another hinged tessellation (see Figure T9). Again, the triangles and the stars are in white, and the background is shaded. Figure T9 on the left shows the tessellation

117

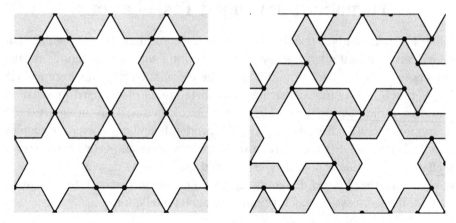

T9: Hinged tessellation of a hexagram and a triangle

at its maximum expansion and on the right shows the tessellation with the triangles rotated clockwise halfway toward the orientation in which no background is visible.

You may have seen a tessellation of the plane that uses octagons and squares. This tiling pattern adorns the floors of some old bathrooms and appears in Figure 12.1. Here we hinge every second vertex of the octagon with the neighboring vertex of the square. Figure T10 on the left shows the tessellation at its maximum expansion; on the right it shows the tessellation with the squares rotated 1/3 of the way clockwise from this first orientation toward an orientation in which no background is visible.

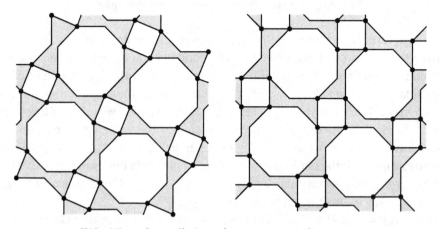

T10: Hinged tessellation of an octagon and a square

CHAPTER 12

With a little Help from my Friends

Some polygons tile the plane directly, and others tile after just a few fortuitous cuts. However, some enlist the assistance of another polygon – they get by with a little help from their friends. Harry Lindgren (1964b) called this technique *completing the tessellation* and surveyed some notable examples. I also discussed this technique in my first book, but this time we will really get high, on the loveliest of puzzles. — *Gonna try!*

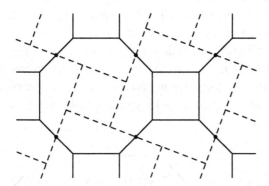

12.1: Superposition of tessellations for an octagon to a square

The octagon and a small square of the same side length together tile the plane, as shown with the solid lines in Figure 12.1. A pair of unequal squares also tiles the plane, as shown with the dashed lines in the figure. The large square is of area equal to that of the octagon, and the small squares are congruent to each other. Superposing these two tessellations, Lindgren (1951) produced the diagram of Figure 12.1, in which I have also shown the symmetry points. The dashed lines in the interior of the octagon indicate its dissection, and the solid lines indicate the corresponding fitting of the pieces into the large square. In this case the "friend" is the little square that helps the octagon tile the plane.

This technique produces the remarkable 5-piece dissection of an octagon to a square that appeared in the anonymous Persian manuscript, *Interlocks of Similar*

119

or Complementary Figures, from approximately 1300. Cambridge mathematician Geoffrey T. Bennett independently discovered – and Henry Dudeney (*Strand*, 1926c) published – this unhingeable dissection. It bettered a 7-piece effort by Dudeney (*Strand*, 1926b), who had already stated in (*Strand*, 1926a) his belief that no such dissection had been previously published. In fact, James Blaikie had earlier posed the problem of dissecting an octagon to a square, for which Henry Martin Taylor (1909) gave an 8-piece solution.

I have managed a 7-piece hingeable dissection of an octagon to a square, based on the 5-piece Persian dissection. In the 5-piece dissection, four of the pieces are identical and surround a small square in both the octagon and the large square. Conveniently, we can hinge those four pieces together. In fact, there is a hinge point to spare, since there are four available hinge points but we need only three of them to hinge four pieces. The challenge is to also hinge the small square. My solution is to view the extra hinge point as a second "friend." Let's use all four of the hinge points and split one of the four identical pieces so that the small square can slip through.

When we split that piece, we force two of the constituent pieces to be identical, so that we can switch their positions. I show such a splitting in Figure 12.2, where the two identical pieces are triangles. Moreover, one triangle is adjacent to the small square inside the octagon in precisely the same way as the other triangle is adjacent to the small square inside the large square. Thus I can merge the small square with the adjacent triangle in the resulting dissection (see Figure 12.3). Two of the hinges abut in both the octagon and the square, yet the pieces in Figure 12.4 are linearly hinged and hing-snug. — *Love at first sight!*

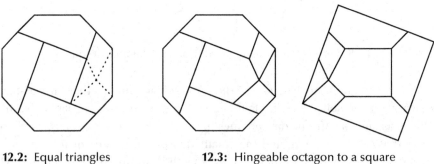

12.2: Equal triangles   **12.3:** Hingeable octagon to a square

12.4: Hinged pieces for an octagon to a square

The technique of completing the tessellation also works for dissecting a do-decagon to a hexagon, since a dodecagon plus two triangles of equal side length tile the plane, and so do a hexagon and two triangles. Lindgren (1964b) credits the resulting 6-piece unhingeable dissection to Los Angeles architect Ernest Irving Freese. We see the superposition of tessellations for this dissection in Figure 12.5, with the shared points of 2-fold rotational symmetry indicated by small dots.

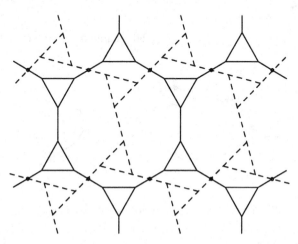

12.5: Superposition of tessellations for a dodecagon to a hexagon

As in the dissection of an octagon to a square, we can hinge together four of the pieces based on these symmetry points. We would like to apply the same tech-niques as in Figures 12.2 and 12.3, but there is a catch. Now there are two small triangles to switch around, rather than one small square. How to juggle too many friends can be a challenging puzzle.

To cope with it, I use the trick in Figure 5.7 of combining the two small triangles in the interior of the dodecagon to make one 60°-rhombus. We can swing the two small triangles in the interior of the hexagon into an area that is also a 60°-rhombus. The dodecagon in Figure 12.6 has dashed edges showing three pieces that bring to-gether the two interior triangles. This is an example of the T(b)-swing technique. I also bring together the two triangles from the outside. Finally, I use the trick from Figure 12.3 to switch the rhombus from one position to the other, indicating the two thin triangles with dotted lines. Again, there is a pair of abutting hinges. Behold the resulting 12-piece hinged dissection in Figures 12.7 and 12.8. — *What do you see?*

The {8/3} and the square seem perfectly suited for completing the tessella-tion. I found a 8-piece unhingeable dissection that uses this technique. Earlier, Lindgren (1964b) had given a lovely 9-piece unhingeable dissection that also uses the technique, and it is that dissection that I modify here.

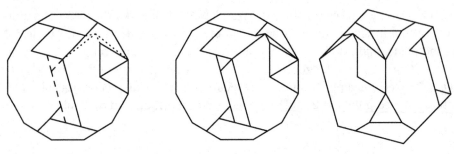

12.6: Tricks, triangles   **12.7:** Hingeable dodecagon to a hexagon

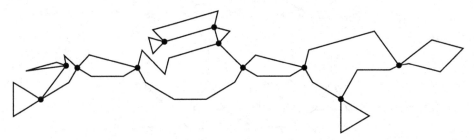

12.8: Hinged pieces for a dodecagon to a hexagon

Lindgren clipped four of the {8/3}'s points and swung them around to form a figure that, along with a square, tiles the plane. Then, in Figure 12.9, he superposed this tessellation with a tessellation of two different squares whose areas are equal to the two figures in the first tessellation.

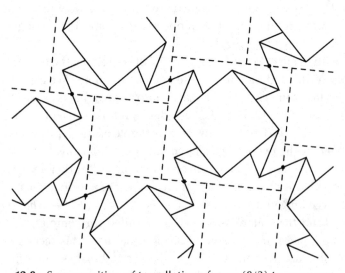

12.9: Superposition of tessellations for an {8/3} to a square

Eight of the pieces in Lindgren's dissection hinge together, and there is once again an extra hinge at our disposal. I use an idea for my 11-piece hingeable dissection (Figure 12.10) that is similar to what I used in the octagon to square. Many variations are available for forming the duplicate pieces, but if you love triangles, as I do, you are out of luck. I chose a concave quadrilateral so that the duplicate pieces are a bit bigger. Several variations are possible for hingings, too, but I have found only one that is not wobbly hinged. I display this hinging, which is hinge-snug, in Figure 12.11. Again, a pair of hinges is abutting. — *What do I do?*

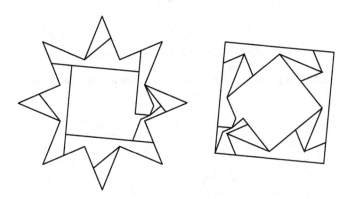

12.10: Hingeable {8/3} to a square

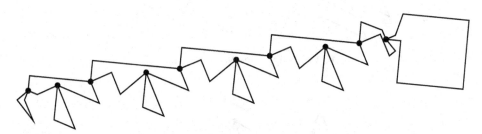

12.11: Hinged pieces for an {8/3} to a square

Bernard Lemaire designed {M̌}, a version of the Maltese Cross in which all line segments in its boundary have a slope whose absolute value is either 1/2 or 2. I found a 5-piece unhingeable dissection to a square, which Pierre Berloquin (*Le Monde,* 1974b) published. I discovered the dissection by adding a small square to the cross and superposing a tessellation of these with a tessellation of large and small squares, using the "completing the tessellation" technique. We see the superposition in Figure 12.12, with the symmetry points also shown.

As in my hingeable dissection of an octagon to a square, I split one of the four identical pieces into four pieces. The two small triangles are the same, and they

123

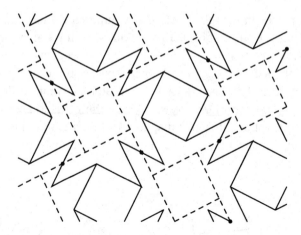

12.12: Superposition of tessellations for a {M̌} to a square

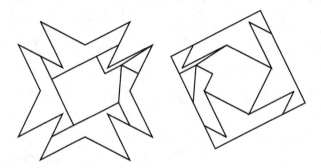

12.13: Hingeable {M̌} to a square

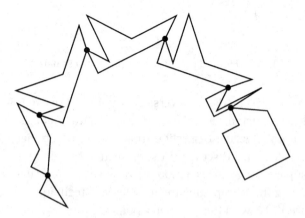

12.14: Hinged pieces for an {M̌} to a square

exchange positions in moving from the cross to the square. As before, I can attach one of them to the square, producing the 7-piece dissection in Figure 12.13 and the hinged pieces (which are hinge-snug) in Figure 12.14. All in a day's work, though the dissection is rather severely wobbly hinged, and a pair of hinges is abutting.
— *How do you feel by the end of the chapter?*

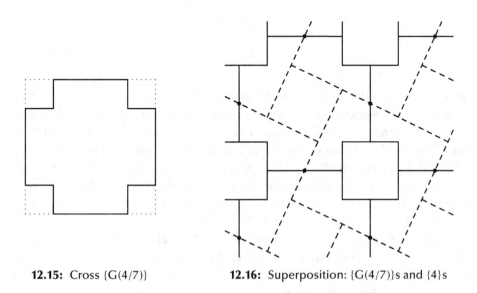

12.15: Cross {G(4/7)}   **12.16:** Superposition: {G(4/7)}s and {4}s

I conclude this chapter with a generalization of a Greek Cross. Take a square and cut four small squares from the corners. Let r be the ratio of the length of the side of the resulting cross to the side length of the original square. I represent the resulting cross as {G(r)}. Thus a {G(1/3)} is the standard Greek Cross. The cross {G(4/7)} stands before us in Figure 12.15. When $1/\phi^2 < r < 1$, where $\phi = (1 + \sqrt{5})/2 \approx 1.618$ is the *golden ratio,* there is a pretty 5-piece unhingeable dissection of a {G(r)} to a square. Completing the tessellation leads to this dissection, as we see in Figure 12.16. For this range of r, we can find a 7-piece hingeable dissection of {G(r)}.

Puzzle 12.1 Find a 7-piece hingeable dissection of {G(4/7)} to a square.

— *Oh, I get by with a little help from my friends.*

125

CHAPTER 13

At Cross Purposes

Commencing with the Greek Cross in Chapter 1, we have witnessed a sublime procession of crosses, including the Latin Cross in Chapter 4, the Cross of Lorraine in Chapter 10, Bernard Lemaire's variation of a Maltese Cross in Chapter 12, and my generalization of the Greek Cross (also in Chapter 12). Crosses manifest themselves in bountiful variety, providing a challenge first for design and then for dissection. Yet especially when we swing around to hingeable dissections, we find that ornamental beauty and hingeability seem to be at cross purposes: the more intricate the design, the more devilish to hinge the dissection.

So when a dissection of a cross turns out to be hingeable, we should feel blessed. Let's not play the skeptic and condemn our dissections for being wobbly. Instead, let us rejoice in their awe-inspiring form. First, let's renew our spirit with a Maltese Cross – a divine shape – and see how we can convert it, swinging as we go.

Henry Dudeney (*Strand,* 1920a) gave a 13-piece hingeable dissection of a Maltese Cross to a square. Unfortunately, the dissection is wobbly hinged. A. E. Hill's remarkable 7-piece dissection, which Dudeney announced in (1926), supplanted that first effort. Regrettably, that remarkable dissection is unhingeable. However, I have found an 8-piece hinged dissection that leads the way in Figures 13.1 and 13.4.

It uses a tessellation approach that is related to Anton Hanegraaf's dissection of a Maltese Cross to a Greek Cross. I superpose tessellations in Figure 13.2, indicating points of 2-fold rotational symmetry by dots. The tessellation element in Figure 13.3

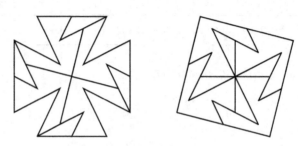

13.1: A Maltese Cross to a square

13.2: Superposition of tessellations for a Maltese Cross to a square

13.3: Maltese element　　　　　**13.4:** Hinges for {M} to a square

has each triangle hinged at its obtuse angle. The dissection has 4-fold rotational symmetry and is translational. This dissection is clearly – indeed, severely – wobbly hinged. It is, however, hinge-snug and grain-preserving.

What could be more of a revelation than to dissect one cross to another? Harry Lindgren (1961) gave a 9-piece unhingeable dissection of a Maltese Cross to a Greek Cross. Bernard Lemaire discovered an 8-piece unhingeable dissection, with four pieces turned over, published by Berloquin (*Le Monde,* 1975a). Anton Hanegraaf gave an 8-piece unhingeable dissection with no pieces turned over.

My 12-piece hingeable dissection marches proudly before us in Figure 13.5. The dissection has 4-fold rotational symmetry and is translational. It also uses a tessellation approach; however, in order to avoid extra line crossings, I use a tessellation that leaves more of the center of the Maltese Cross open. We enjoy an unobscured

127

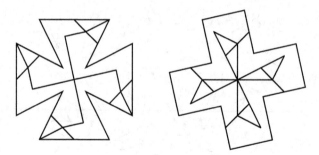

13.5: A Maltese Cross to a Greek Cross

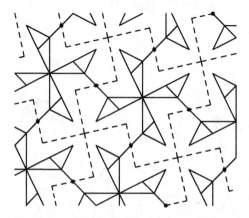

13.6: Superposition of tessellations for a Maltese Cross to a Greek Cross

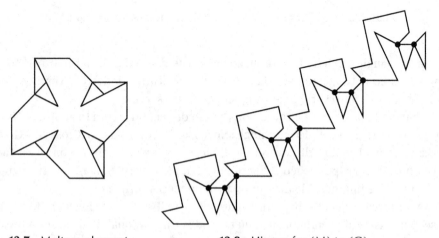

13.7: Maltese element

13.8: Hinges for {M} to {G}

view of the tessellations in Figure 13.6, with the dots indicating points of 2-fold rotational symmetry, and the tessellation element is in Figure 13.7. A set of hinged pieces swing wobbly past in Figure 13.8. This dissection is also hinge-snug and grain-preserving.

In 1999, Bernard Lemaire sent me a treasure chest of dissections of crosses to squares, including one in Figure 13.9 that dominates the scene. Each arm of the

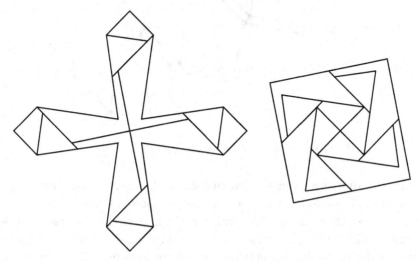

13.9: A wide-tie cross $\hat{M}_2$ to a square

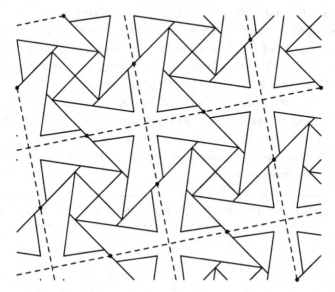

13.10: Superposition of tessellations for a wide-tie cross to a square

129

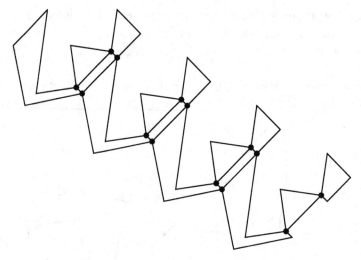

13.11: Hinges for a wide-tie cross to a square

cross has a width that is twice the distance across the "neck" of the cross, and the height from top to bottom is ten times the distance across the neck. Furthermore, the angles at the extreme ends of the arms are 90°. Each cut that intersects the long side of an arm bisects the side, and the four triangles created are isosceles triangles. We can derive the dissection by superposing tessellations, as in Figure 13.10.

Bernard did not realize it, but the dissection is hingeable, as we see in Figure 13.11. In fact, it is a member of a family of crosses, each of which has a 12-piece hingeable dissection to a square. These dissections are wobbly hinged, hinge-snug, and grain-preserving. Let $p \geq 1$ be the ratio of the width of an arm to the distance across the neck. Then the ratio of the height from top to bottom over the distance across the neck is $4p + 2$. The side length of the square is $\sqrt{(2p+1)^2 + 1}$ times the distance across the neck. Bernard's cross corresponds to the case of $p = 2$. Since each arm of the cross looks like a rather wide necktie, I call this the family of wide-tie crosses. I designate a wide-tie cross of ratio p by $\hat{M}_p$. When $p = 1$, we need only eight pieces for a hinged dissection. Let's find this cross and bring a lovely end to our glorious procession!

Puzzle 13.1   Find 8-piece hingeable dissection of a $\hat{M}_1$ to a square.

Curious Case, part 4: A Dudeneyan Slip?

A year after he first posed the puzzle of dissecting an equilateral triangle into a square, Henry Dudeney (*London*, 1903a) reframed his problem as "The Haberdasher's Puzzle" and included it in the final installment of his puzzle series "The Canterbury Puzzles":

> Many attempts were made to induce the Haberdasher, who was of the party, to propound a puzzle of some kind, but for a long time without success. At last, at one of the Pilgrims' stopping-places, he said that he would show them something that would "put their brains into a twist like unto a bell-rope." As a matter of fact, he was really playing off a practical joke on the company, for he was quite ignorant of any answer to the puzzle that he set them. He produced a piece of cloth in the shape of a perfect equilateral triangle, as shown in the illustration, and said, "Be there any among ye full wise in the true cutting of cloth. I trow not. Every man to his trade, and the scholar may learn from the varlet and the wise man from the fool. Show me then, if ye can, in what manner this piece of cloth may be cut into four several pieces that may be put together to make a perfect square."
>
> Now some of the more learned of the company found a way of doing it in five pieces, but not in four. But when they pressed the Haberdasher for the correct answer he was forced to admit, after much beating about the bush, that he knew of no way of doing it in any number of pieces. "By Saint Francis," saith he, "any knave can make a riddle methinks, but it is for them that may to read it aright." For this he narrowly escaped a sound beating. But the curious point of the puzzle is that we have found that the feat may really be performed in so few as four pieces, and without turning over any piece when placing them together. The method of doing this is subtle, but we think the reader will find the problem a very interesting one.

The first installment of "The Canterbury Puzzles" had appeared in the January 1902 issue of *The London Magazine,* so that Dudeney would have already had this series on his mind when Charles McElroy submitted his 4-piece solution to the triangle-to-square problem in the *Weekly Dispatch.* Thus it is remarkable that Dudeney framed the puzzle in this rather curious fashion in "The Canterbury Puzzles" – namely, of the haberdasher challenging the company to find the 4-piece solution when he doesn't know such a solution himself.

Was this a hidden allusion to what may have happened when Dudeney posed the original version of the puzzle in the *Weekly Dispatch*? In his *Dispatch* column of April 20, 1902, Dudeney implied that he had intended a 4-piece solution. But perhaps he had known only of a 5-piece solution.

Dudeney presented the puzzles in "The Canterbury Puzzles" in a highly styl-ized form, not unlike a dream. Can techniques appropriate to the Freudian inter-pretation of dreams be useful here? In applying Sigmund Freud's theory of the interpretation of dreams to literary criticism, Frederick Hoffman (1945) empha-sized that the thoughts providing the basis for the dream are allowed to come through in disguised form. Dream elements of high psychic significance are dis-torted to mask their meaning from the conscious mind. The actual dream is highly condensed. Every element of the actual dream enjoys a manifold representation in the dream thoughts. Words are in general treated as things within the dream, and thus undergo similar displacements and substitutions. The resulting play on words is not unlike that of children. If a proper name resists visual represen-tation, it will be distorted considerably, or even replaced by rather far-fetched references.

Let's attempt to match phrases in the puzzle statement with real and hypoth-esized circumstances.

"Many attempts were made ... but for a long time without success"
Henry Dudeney was already 40 by the time he began to achieve any real success with his own puzzle column.

"he would show them something that would put their brains into a twist"
Dudeney had earlier referred to the "inherent difficulty" of the puzzle.

"he was quite ignorant of any answer to the puzzle that he set them"
Presumably, Dudeney knew a 5-piece solution, but this is unsatisfactory when a 4-piece solution is possible.

"he was really playing off a practical joke on the company"
The worst sort of joke on himself is for Dudeney to have discussed the inherent difficulty of an inferior solution.

"the scholar may learn from the varlet and the wise man from the fool"
With little formal schooling, Dudeney could not penetrate serious mathe-matical circles, even if some of his puzzles were of mathematical interest.

"Now some of the more learned of the company found a way of doing it in five pieces, but not in four"
Five-piece solutions had been found by some of Dudeney's readers.

"But when they pressed the Haberdasher for the correct answer"
The pressure of producing a weekly column without mistakes must have seemed substantial to Dudeney.

"after much beating about the bush"
Dudeney delayed publishing the solution in the *Weekly Dispatch* for an extra two weeks.

"For this he narrowly escaped a sound beating"
    Self-censure of Dudeney translated into feared public censure?

'Haber-dashery' – 'Henry Dudeney'?

Did Dudeney mock himself as a haberdasher? At that time, a haberdasher was a dealer in small wares or notions. Are not Dudeney's puzzles similar to small wares and notions? Maurice Hussey (1968) and Francis King and Bruce Steele (1969) point out that, in the *Canterbury Tales,* Geoffrey Chaucer treated the five guildsmen – of whom the haberdasher was one – with irony. The guildsmen were the new rich. Chaucer built them up only to deflate them by showing the true basis of their pomp: sufficient wealth to qualify for aldermanic office and a desire to satisfy the vanity of their wives. Not members of the larger trade guilds, the guildsmen were all lumped together in a lesser guild. They had no tales and disappeared after the Prologue.

Certainly, Dudeney's position in British society was not particularly impressive: he was a journalist, but specifically someone who wrote puzzle columns. He promoted his columns by awarding prizes for the best solutions. Although Dudeney was a member of the Authors' Club, he could not be compared with more illustrious members of this "guild." For example, his name did not appear among the notables in the *Times* newspaper articles describing the club's dinners, nor was he ever listed as chairman of these in a list that commenced with the club's reconstruction in 1908.

Is Dudeney's tale of the Haberdasher a camouflaged description of himself, in which he alludes to his own missteps and shortcomings? Dare we call his tale a "Dudeneyan slip"?

CHAPTER 14

Handling the Curves

In the game of baseball, a player on one team pitches a ball in front of a player on the other team, who tries to hit the ball with a long round stick called a bat. It should be easy for the batter to hit the ball, except that the pitcher can throw the ball fast or with a spin that makes it "slide," or wobble, or curve. Handling the fast balls would not be so bad if the batter did not also have to handle the curves.

So it is time for me to throw you, the sporting reader, some curved figures. Their dissections are not so hard if only we do not also require them to be hinged. In fact I do not know if a hinged dissection of two curved figures exists whenever an unrestricted dissection of them exists. Some of the simpler dissections reach back to Leonardo da Vinci in the Renaissance. And some have just appeared in our own renaissance of hinged dissections.

As James McCabe discussed in (1972), Leonardo da Vinci was fascinated by the problem of calculating the area of curved figures. Leonardo recognized two techniques: One is to apply the Pythagorean theorem to semicircles whose diameters correspond to the legs of a right triangle. The second technique is to dissect geometric figures whose concave portions of their circular boundaries exactly match their convex portions. Thus Leonardo identified some simple dissections of curvilinear figures to squares. In (1973, *Codex A,* 6c and 12a), he gave a dissection of a "pendulum" to a square, on the left in Figure 14.1. The pendulum consists of two "falcatas," which are triangles with two curved sides. In (1973, *Codex A,* 5a), Leonardo showed how to convert a falcata with two matching curved sides to an isosceles right triangle and went on to explain the method that gives the dissection on the left in the figure.

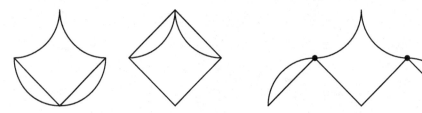

14.1: Hinged dissection of a pendulum to a square

Another simple dissection that Leonardo (1973, *Codex A,* 6b) gave is of an "axe-head" to a square, on the left in Figure 14.2. Both dissections are rather simple and require just three pieces. I have found no evidence in Leonardo's manuscripts that he realized the dissections are hingeable, though he would certainly have loved to see his pendulum swing. Herbert Wills (1985) put this spin on them, recognizing that they are hingeable. He also gave hinged dissections of these figures to a (2×1)-rectangle.

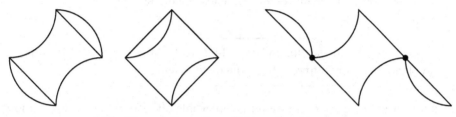

14.2: Hinged dissection of an axe-head to a square

Wills discussed Leonardo's investigations into converting a curvilinear figure into a rectangular figure. And he made it into the big leagues by pitching some beautiful curves. Wills chose as an illustration a certain curved figure that he termed a "motif." Leonardo (1973, *Codex A,* 5b) drew it individually and as part of a design containing eight motifs. Leonardo (1973, *Codex A,* 6a) gave it again individually and as part of a design containing sixteen motifs.

Although it is apparent from context that Leonardo recognized the connection between his motif and a rectangle, he did not give an explicit conversion from his motif to either a rectangle or square. Wills gave such a dissection of the motif to a (2×1)-rectangle. He performed the dissection in two steps, with each step being hingeable. However, the combination of both steps together does not give a completely valid hinging. I round up a 6-piece variation of Wills's dissection that is completely hingeable in Figure 14.3; the hinged pieces turn up in Figure 14.4.

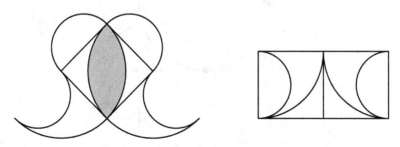

14.3: Hingeable dissection of a Leonardo's motif to a rectangle

135

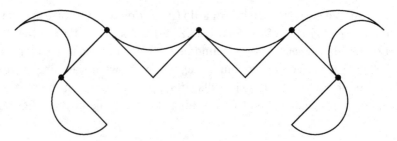

14.4: Hinged pieces for a Leonardo's motif to a rectangle

Puzzle 14.1 The dissection in Figure 14.3 can be hinged in more than one way. Find a hinging different from the one in Figure 14.4.

Why stop with a rectangle? Can we find a hingeable dissection of Leonardo's motif to a square? My 10-piece hingeable dissection in Figure 14.5 squares up the round

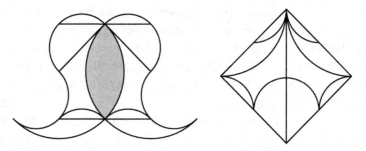

14.5: Hingeable dissection of a Leonardo's motif to a square

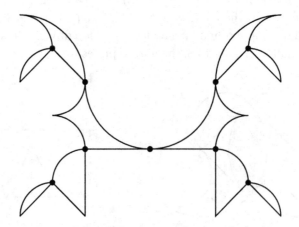

14.6: Hinged pieces for a Leonardo's motif to a square

edges. The hinged pieces in Figure 14.6 suggest a pair of antlers – from some strange mathematical beast! Now we have touched all the bases, and we are *swinging!*

Puzzle 14.2   The dissection in Figure 14.5 can also be hinged in more than one way. Find a hinging different from the one in Figure 14.6.

In the spring of 1997, Gavin Theobald considered a pretty "globular cross," in which the combined length of the inward curves equals the length of the outward curves (Figure 14.7). Gavin found nifty unhingeable dissections to a square (8 pieces, with 2 turned over) and to a Greek Cross (10 pieces). Each arc of the globular cross has a radius of 2.5 units. It is not too hard to find a 13-piece hingeable dissection, and I found a 12-piece hingeable dissection. But then Gavin smacked this one right out of the ball park by finding two different 11-piece hingeable dissections, one of which I include in Figure 14.8; the linearly hinged pieces march past in Figure 14.9.

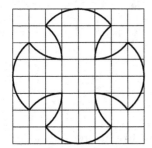

14.7: Globular Greek Cross   **14.8:** Theobald's globular Greek Cross to a square

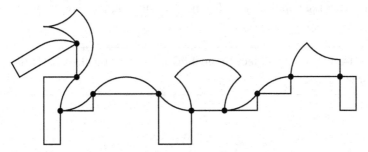

14.9: Hinged pieces for a globular Greek Cross to a square

In response to the globular Greek Cross, I designed the globular Latin Cross (Figure 14.10). I found an unhingeable dissection to a square in 8 pieces, and Gavin then found an unhingeable dissection to a Greek Cross in 10 pieces. Now I have circled back to find the 16-piece hingeable dissection to a square in Figure 14.11. The

137

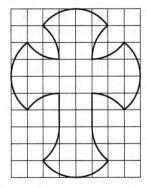

14.10: Globular Latin

14.11: Globular Latin Cross to a square

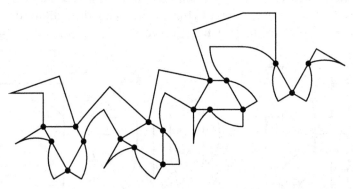

14.12: Hinged pieces for a globular Latin Cross to a square

hinged pieces are in Figure 14.12. Although the dissection is wobbly hinged, it has a nice feature: three small cycles of hinges in the hinging. Not a bad rotation, for throwing out curves.

Puzzle 14.3 Find a 16-piece hingeable dissection of a globular Latin Cross to a Greek Cross.

CHAPTER 15

And Also Starring

For overall utility in hingeable dissections, squares, triangles, and hexagons have earned their spot in the limelight. Yet for elegance and glamour, stars deserve top billing. These celestials normally rely on roles designed with their special properties in mind – in particular their internal structure, which consists of rhombuses and halves of rhombuses (see Figure 15.1). Harry Lindgren (1964b) showed how to use cuts along the sides or diagonals of these rhombuses to produce many beautiful dissections. In my first book, I called them *auspicious* dissections and listed the special trigonometric relationships on which they depend. Now we will see how to extend this rhombic technique to achieve a grace in hinged motion that is truly stellar.

The first member of our star-studded extravaganza is a lovely dissection of a pentagram to a decagon. Lindgren gave a 6-piece unhingeable dissection based on internal structures of both the pentagram and decagon, which each contain five 36°-rhombuses and five 72°-rhombuses. Given the rhombic structure of the pentagram in Figure 15.1, can readers deduce a related structure for the decagon? We need only minor changes from Lindgren's to create the 7-piece hingeable dissection in Figure 15.2, but hinging the pieces is tricky. I give a linear hinging in Figure 15.3. There is another way to hinge the dissection, one that is not linear. Can you spot it?

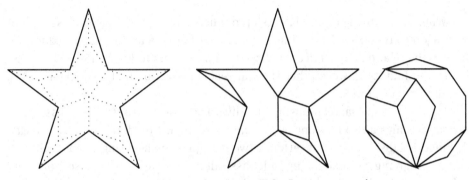

15.1: Pentagram structure   **15.2:** Pentagram to decagon

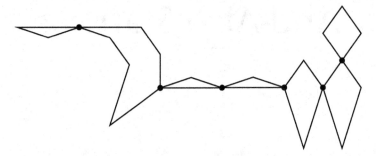

15.3: Hinged pieces for a pentagram to a decagon

We reserve the greatest acclaim for the stars that come in families. Lindgren identified a special trigonometric relationship between a $\{(4n+2)/(2n-1)\}$ and a $\{(2n+1)/n\}$, for any positive integer n. It leads to a $(4n+2)$-piece hingeable dissection of a $\{(4n+2)/(2n-1)\}$ to two $\{(2n+1)/n\}$s. Don't let the mystique of this family intimidate you. Just substitute a number, such as 2, in place of n, and simplify numerators and denominators to determine that there is a 10-piece hingeable dissection of a $\{10/3\}$ to two $\{5/2\}$s. In Figure 15.4, I show the hinged dissection of a $\{14/5\}$ to two $\{7/3\}$s, corresponding to $n = 3$.

15.4: Hinged $\{14/5\}$ to two $\{7/3\}$s

When $n = 1$, the special relationship predicts a 6-piece hingeable dissection of a hexagon to two triangles. In this case, neither figure is a star, but that does not tarnish the relationship. Indeed, a dissection analogous to Figure 15.4 exists. Can the reader find it? However, there is a hingeable dissection that has only 4 pieces. It is hidden in Figure 5.6.

Whenever a special relationship identifies a one-into-two auspicious dissection, it also identifies a two-into-one auspicious dissection. Not all of these lead to easily hingeable dissections, but certainly two hexagons to one triangle is not difficult. Harry Lindgren gave such a hingeable dissection, in 6 pieces. His dissection must use two abutting hinges, which we avoid with the dissection in Figure 15.5.

140

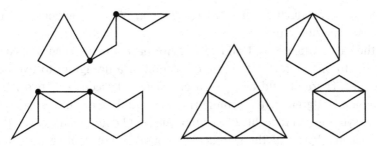

15.5: Hinged dissection of two hexagons to a triangle

The next family of stars makes an even bigger splash. It relies on the special relationship between a $\{(2n+2)/n\}$ and a $\{2n+2\}$, for any positive integer n. For the case when $n = 3$, Lindgren gave the 9-piece dissection of an $\{8/3\}$ to two octagons (on the right in Figure 15.6), which he did not identify as hingeable. Remarkably, this dissection appeared as part of a larger construction in the anonymous Persian manuscript, *Interlocks of Similar or Complementary Figures,* from around 1300. The text also includes a description of a $\{10/4\}$ to two decagons. Alpay Özdural (2000) conjectured that the designer of the construction was an artisan.

15.6: Hinged $\{8/3\}$ to two octagons

Since the relationship yields a one-to-two dissection, we can turn it around to identify an auspicious dissection of two $\{8/3\}$s to an octagon. Harry Lindgren gave a corresponding 13-piece dissection (on the right in Figure 15.7), but again he did not mention that it is hingeable. The approach illustrated in Figures 15.6 and 15.7

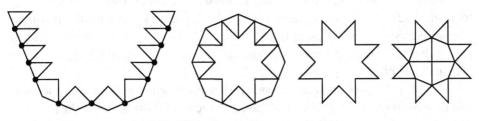

15.7: Hinged octagon to two $\{8/3\}$s

141

works for any n, producing $(2n + 3)$- and $(3n + 4)$-piece dissections, respectively. For $n < 3$, we can do better.

For the cases when $n = 1$ and $n = 2$, we have already seen a dissection of two squares to one (Figure 1.2) and a dissection of a hexagram to two hexagons (Figure 3.25). There should also be an auspicious dissection of two hexagrams to a hexagon. Lindgren gave a symmetric 6-piece unhingeable dissection for this case. For a symmetric hingeable dissection, the best I could manage was 10 pieces. However, when I let go of the symmetry, I found a 9-piece hingeable dissection (Figures 15.8 and 15.9). We have not lost all symmetry, since all of the cuts in the hexagram on the left are in the hexagram on the right.

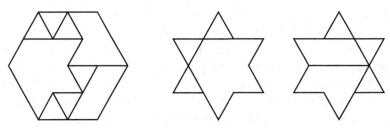

15.8: Hingeable dissection of two hexagrams to a hexagon

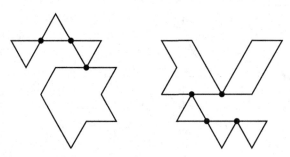

15.9: Hinged pieces for two hexagrams to a hexagon

You may feel that most of the dissections in this family rely on simple techniques. On the contrary, the family achieves true superstardom for odd values of n. For these values of n, there are auspicious dissections of a single $\{(2n + 2)/n\}$ to a single $\{2n + 2\}$. For the case of $n = 3$, Lindgren gave a beautiful 6-piece unhingeable dissection of an $\{8/3\}$ to an octagon. His key insight is that we can form an eighth of an octagon from an isosceles right triangle and a smaller right triangle with an angle of $22\frac{1}{2}°$.

I found an 11-piece hingeable dissection loosely based on Lindgren's dissection, and then Anton Hanegraaf found a 9-piece dissection in which one of the pieces is flipped over. From that, I then found the irregular 10-piece hinged dissection that

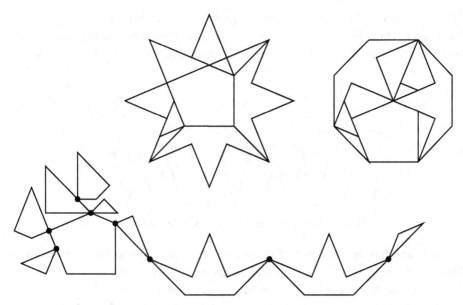

15.10: Hinged dissection of an {8/3} to an octagon

sits a bit awkwardly in Figure 15.10. A useful insight is that cutting a smaller isosceles right triangle off of an isosceles right triangle allows us to reform the isosceles right triangle in a different orientation.

For the case of $n = 5$, Lindgren gave a clever 11-piece unhingeable dissection of a {12/5} to a dodecagon, which I improved to 10 pieces in (1972a). In fact, both are decidedly unhingeable, so that the best hingeable dissection that I have found (Figure 15.11) uses 18 pieces. I take advantage of the fact that the twelve points of

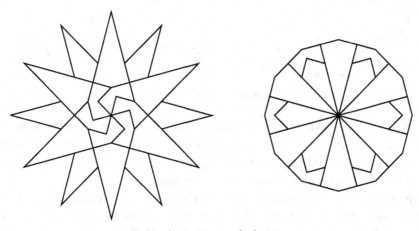

15.11: A {12/5} to a dodecagon

143

the {12/5} sum to 360°, allowing me to group them together in the center of the dodecagon. Then the interior of the star neatly fills in around the border of the dodecagon. The hinged pieces in Figure 15.12 are wobbly hinged, but the dissection is hinge-snug.

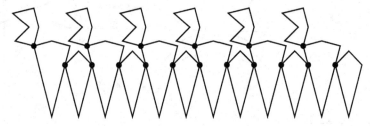

15.12: Hinged pieces for a {12/5} to a dodecagon

The next family of stars is a show-stopper. Harry Lindgren reported on the family of dissections of a {(2n + 4)/2} to an {n + 2}, for every positive integer n. Corresponding to the case $n = 3$, we have already seen a hingeable dissection of a {12/2} to a {6} in Figure 10.21. Let's next examine the cases for $n = 1$ and $n = 2$.

A 6-piece unhingeable dissection of a hexagram to a triangle appeared in the 700-year-old *Interlocks* manuscript. Geoffrey Mott-Smith (1946) found several 5-piece dissections, one of which is on the right in Figure 15.13. Neither Mott-Smith nor Lindgren, who also gave this dissection, identified it as hingeable. Yet it is hingeable in five different ways.

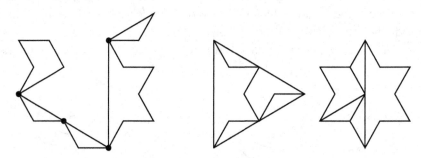

15.13: Hinged dissection of an {6/2} to a triangle

Lindgren (1961) also gave a different dissection of a hexagram to a triangle, which radiates such beauty on the right in Figure 15.14. As with the previous dissection, he did not identify it as hingeable, but it also is hingeable in five different ways. I have shown the most symmetric of the hingings.

For $n = 2$, the relationship indicates an auspicious dissection of an {8/2} to a square. Lindgren (1964b) gave an 8-piece unhingeable dissection, which (unbeknownst to him) had also appeared in the *Interlocks* manuscript. I improved this

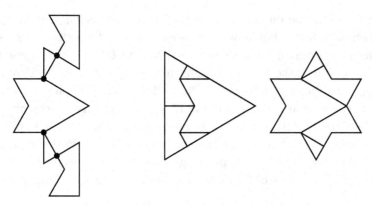

15.14: Another hinged dissection of an {6/2} to a triangle

dissection to a 7-piece unhingeable dissection in (1972d) and later found two more 7-piece unhingeable dissections. I struggled to get a 12-piece hingeable dissection by modifying one of those. When Anton Hanegraaf saw it, he discovered a nice way to reduce the number of pieces by one; see Figure 15.15. His modification is clever, because he uses some cuts that do not fall on sides or diagonals of rhombuses. These cuts create two small isosceles right triangles that he swaps around in Figure 15.16.

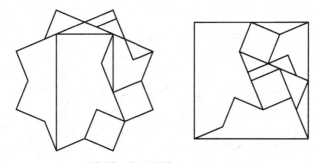

15.15: An {8/2} to a square

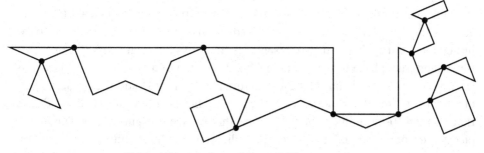

15.16: Hinged pieces for an {8/2} to a square

145

Sometimes one star outshines the rest of the family, as happens with the family based on the relationship between a $\{(4n + 4)/2\}$ and a $\{(2n + 2)/n\}$. We have already seen a dissection for the case $n = 1$ in Figure 15.15. The case for $n = 2$ yields an auspicious dissection for a $\{12/2\}$ to a hexagram. Harry Lindgren gave an elegant 9-piece unhingeable dissection. Related to it is the simpler 12-piece dissection that shines so lustrously in Figure 15.17. Once we cut six 60°-rhombuses from the outside of the $\{12/2\}$, we can cut the remainder into six congruent pieces, each forming a point of the hexagram. The rhombuses then fill in the interior of the hexagram. With its hinged pieces as in Figure 15.18, this dissection is hinge-snug. Surely, Lindgren was aware of the intuition behind this very symmetrical dissection.

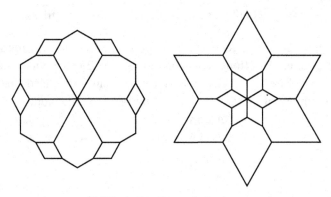

15.17: A $\{12/2\}$ to a hexagram

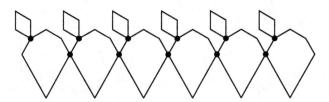

15.18: Hinged pieces for a $\{12/2\}$ to a hexagram

Our next family of stars relies on a relationship between a $\{(4n + 2)/(n + 1)\}$ and a $\{(4n+2)\}$. With $n = 1$, we have already seen dissections of a hexagram to two hexagons (Figure 3.25) and two hexagrams to a hexagon (Figure 15.8). For $n = 2$, Harry Lindgren gave a 6-piece unhingeable dissection of a $\{10/3\}$ to two decagons. His dissection depends on internal structures of the $\{10/3\}$ and the decagon, each of which contains 36°-rhombuses and 72°-rhombuses. I have adapted his dissection into the 10-piece hingeable dissection that we see in Figure 15.19. The hinged pieces are wobbly hinged, but at least the dissection is hinge-snug.

146

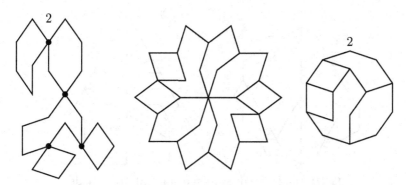

15.19: Hinged dissection of a {10/3} to two decagons

Our last family of stars relies on the relationship between a $\{(4n + 2)/(n + 1)\}$ and a $\{(4n + 2)/(2n)\}$. Alert readers will realize that there is not much to do when $n = 1$. For the case when $n = 2$, Lindgren gave an 11-piece unhingeable dissection of a {10/4} to a {10/3}. He cut the points off of the {10/4}, leaving a decagon, and then arranged the points differently around the decagon to give the {10/3}. I have adapted his dissection into the 21-piece wobbly-hinged dissection in Figure 15.20 by cutting each point of the {10/4} into two pieces.

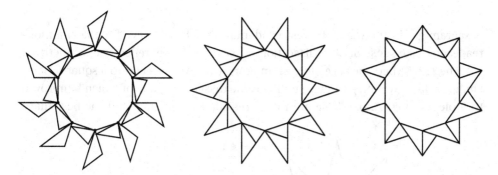

15.20: Hinged dissection of a {10/4} to a {10/3}

Let's conclude our star-studded review with a few isolated dissections that can carry the show by themselves. Lindgren gave a remarkable 6-piece unhingeable dissection of a {12/2} to a triangle. It seems impossible to find a hingeable dissection with so few pieces. Indeed, it is not easy to find any hingeable dissection at all, so that the 16-piece hingeable dissection in Figures 15.21 and 15.22 begins to look rather good. I started with the dissection of a {12/2} to a hexagram in Figure 15.17 and then attempted to apply the dissection of a hexagram to a triangle in Figure 15.13. The latter must be modified, because some of the points of the

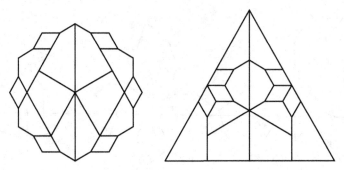

15.21: Hingeable dissection of a {12/2} to a triangle

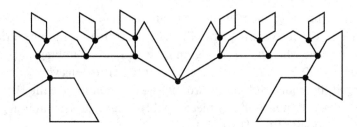

15.22: Hinged pieces for a {12/2} to a triangle

hexagram get in the way. The rest is adaptation, which I leave for the ambitious reader to make sense of. The dissection possesses a lovely reflection symmetry.

The {12/5} becomes even more luminous when we dissect it to a square. Harry Lindgren gave two nifty 10-piece unhingeable dissections, one of which I improved to 9 pieces in (1972a). These are rather resistant to hinging, and the best that I

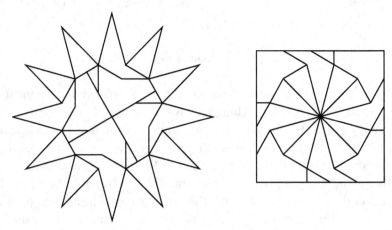

15.23: A {12/5} to a square

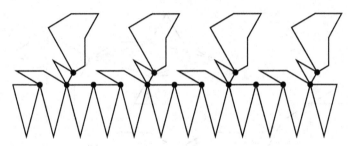

15.24: Hinged pieces for a {12/5} to a square

have found is the 20-piece wobbly-hinged dissection that glows so brightly in Figures 15.23 and 15.24. As in Figure 15.11, I take advantage of the fact that the twelve points of the {12/5} sum to 360°, and I group them together in the center of the square. The interior of the star neatly fills in the corners of the square.

There is always an upstart who presumes to be a star. The internal structures of a Latin Cross and a dodecagon indicate that they have an auspicious dissection. Whether this is enough to admit them to our cavalcade of stars is not clear, but they sneaked in anyway. Lindgren (1962) gave a neat 7-piece unhingeable dissection of a Latin Cross to a dodecagon. Based on the tessellation technique, his dissection seems to resist transformation into a hinged dissection. I found a 12-piece hingeable dissection, which Anton Hanegraaf improved to the 10-piece dissection shown in Figures 15.25 and 15.26. We can view the Latin Cross as the union of six equal squares. If we inscribe the largest possible square in a dodecagon, then this square is equivalent to four of the six squares. The four trapezoids remaining after removing the largest possible inscribed square make up the other two of the six squares.

The 10-piece dissection uses tricks from both of us. Anton's trick is to form the edges of the cross using just 5 pieces, with one of the six squares split to supply

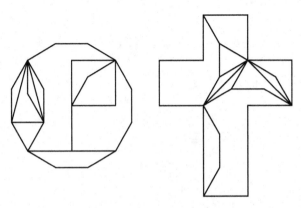

15.25: A Latin Cross to a dodecagon

149

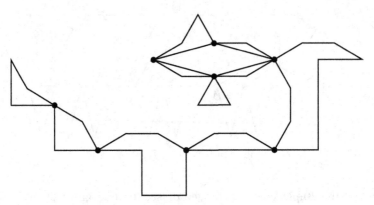

15.26: Hinged pieces for a Latin Cross to a dodecagon

four of the edges. My trick is to split two 30°-rhombuses along their long diagonals, which allows the rhombuses to rearrange from the dodecagon to the cross.
One notable feature is that three pieces attach at one hinge and three at another.
More interesting is that we can cyclicly hinge the pieces that result from splitting
the two rhombuses. If we flap the pieces of this cycle against each other, we create the illusion of clapping hands. Let's use this mechanical applause to gracefully
cover our exit from the chapter.

Turnabout 4: Hinge-Convertible Polyominoes

An n-omino is a figure formed by attaching n squares side-to-side. Not counting reflections, there are five different tetrominoes ($n = 4$) and twelve different pentominoes ($n = 5$). Solomon Golomb (1994) studied these figures extensively, calling them *polyominoes*. In 1997, David Eppstein (at the University of California at Irvine) posed the problem of designing, for any given n, a hinged assemblage that can form any n-omino. David found a way to hinge four square pieces to form any of the five tetrominoes.

For $n \geq 5$, hinging squares does not seem to work. David proposed a way to dissect any n-omino into a cycle of $4n$ smaller squares and then cut each smaller square into a set of 12 pieces that hinge appropriately. Erich Friedman (Stetson University, Florida) refined and simplified these ideas by replacing each of the n squares with two isosceles right triangles, giving a hinged assemblage of $2n$ pieces that form a cycle.

When we pack the cycle into an n-omino, the long sides of the triangles identify diagonals of the n squares that form a "skeleton." Each n-omino has such a skeleton, which guarantees a cycle packing. The cycle of isosceles triangles is in the center of Figure T11, flanked by two pentominoes. Dashed edges indicate the skeleton. This solution is not optimal because we can save two pieces by merging pieces 9, 10, and 1 together.

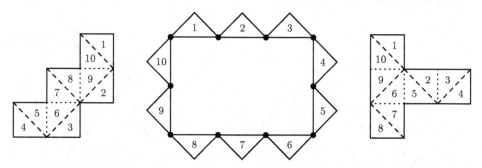

T11: W-pentomino to a T-pentomino using the cycle of isosceles right triangles

Analogous to polyominoes are polyiamonds and polyhexes, which are formed from triangles and hexagons, respectively. Erik and Martin Demaine joined David and Erich in (1999) in explaining how to design a hinge assemblage that would form any n-iamond and another to form any n-hex. They defined a "k-regular" to be a triangle if $k = 3$, a square if $k = 4$, and a hexagon if $k = 6$. They showed that, for any given n and k, there is a $\lceil k/2 \rceil (n - 1)$-piece hinged assemblage that forms any "$n \times k$-regular."

Soon afterwards, David posed the problem of finding, for any given n, one hinged assemblage that would form any n-omino and any n-iamond. He showed how to adapt the 4-piece dissection of a triangle to a square to give a 9-piece cyclicly hinged dissection, so that combining n cycles gives the desired assemblage. He posed similar problems of any n-omino to any n-hex and also any n-hex to any n-omino.

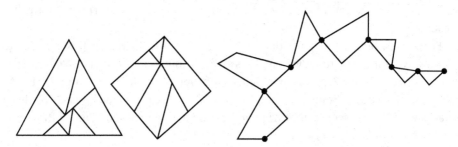

T12: Suitable square to triangle   **T13:** Extendible path hinging

I then found a 7-piece linearly hinged dissection that is sufficient for any n-omino to any n-iamond, as well as a 10-piece linearly hinged dissection suitable for any n-omino to any n-hex, and another suitable for any n-hex to any n-iamond. In the triangle to square (Figures T12 and T13), the points at which to attach additional paths are the "dangling" hinges.

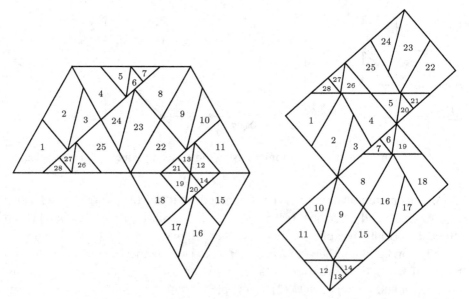

T14: Forming a tetriamond   **T15:** Forming a tetromino

Figures T14 and T15 show how to form a tetriamond and a tetromino from a path consisting of four copies of the extendible path. The pieces are numbered by their order in the path. For $n \geq 4$, we can merge the last 7 pieces in the path. In our examples, these are pieces 22 through 28.

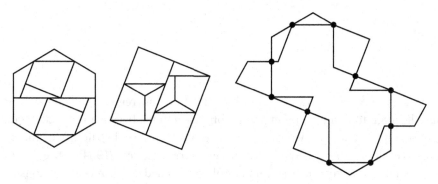

T16: Suitable hexagon to square   **T17:** Extendible cycle hinging

The trick in such extendible hingings is to have a hinge at the midpoint or an endpoint of every side of the square, triangle, or hexagon so that we can switch from filling in one square, triangle, or hexagon to filling in the next in the n-omino, n-iamond, or n-hex, respectively. The cycle for the hexagon to square (Figures T16 and T17) splits at any point to attach additional split cycles.

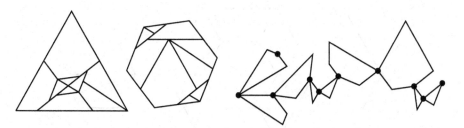

T18: Suitable hexagon to triangle   **T19:** Extendible path hinging

Figure T16 comes from an 8-piece TT2 strip dissection that produces hinge points at the midpoints of four of the six hexagon edges and at an endpoint of every side of the square. Two additional cuts produce hinge points at the midpoints of the remaining sides of the hexagon. Figure T18 is a suitable dissection of a hexagon to a triangle. It comes from a 6-piece TT2 strip dissection with three extra cuts to produce additional hinge points and make possible the extendible path (Figure T19).

153

CHAPTER 16

Four of a Kind

It is an easy task, in general, to find dissections of four regular polygons to one. I would hesitate to include any in a book on dissections, because those dissections are normally child's play. We can base them on the internal structure of the polygons and easily predict the number of pieces. Harry Lindgren (1964b) observed that, for any regular polygon $\{p\}$, the number of pieces used is p if p is even and is $p+1$ if p is odd.

However, Lindgren's simple dissections are not hingeable. Holding out for hingeability raises the stakes considerably. It's then no longer a kid's game like "Go Fish," so you had better turn in your childish smirks for poker faces. Okay, okay … four hexagons to one is an easy case, for which we find the 7-piece hinge-snug dissection winning hands down in Figure 16.1. That's just one piece more than Lindgren's unhingeable dissection. But don't expect dealer's luck all the time.

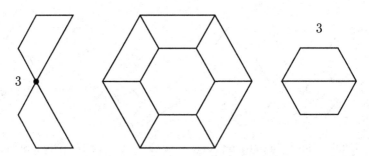

16.1: Hinged dissection of four hexagons to one

If we allow stars as well as polygons when we try to draw four of a kind, things can get tough. Looking at hexagrams, Robert Reid beat the odds by finding an 11-piece unhingeable dissection. I give a 16-piece hingeable dissection in Figure 16.2, with the hinged pieces in Figure 16.3.

Lindgren (1964b) dealt with four pentagons by giving a simple 6-piece unhingeable dissection. My 8-piece hinged dissection steals the hand in Figure 16.4. To use as few as 8 pieces, I suspect that it is necessary to cut each of the pentagons. I have

154

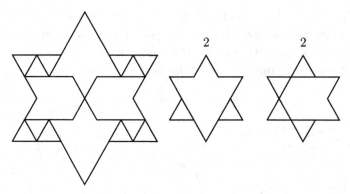

16.2: Hingeable dissection of four {6/2}s to one

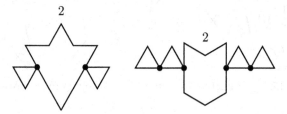

16.3: Hinged pieces for four {6/2}s to one

labeled the pieces in the large pentagon so that you can see how they match up. Then we can easily see how the rhombuses must hinge with the remaining pieces. Can you spot the simple 6-piece unhingeable dissection that results by merging pieces?

Turning to four heptagons, Lindgren (1953) laid down an unhingeable dissection using just 8 pieces. However, coming up with a good hingeable dissection is like drawing to an inside straight. My hingeable dissection in Figure 16.5 uses 12 pieces. I cut each small heptagon in the same way as one other and then label the

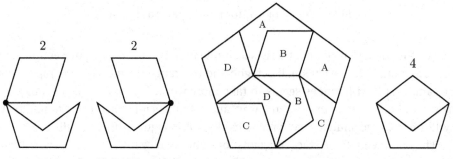

16.4: Hinged dissection of four pentagons to one

pieces in the large heptagon so that you can see how they hinge together. If you examine the hinging in Figure 16.6, you will see that the two heptagons from each identically cut pair hinge differently! This dissection simplifies to an 8-piece un-hingeable dissection by merging four pairs of pieces together.

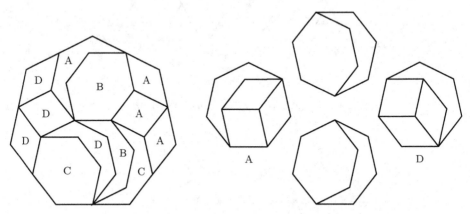

16.5: Hingeable dissection of four heptagons to one

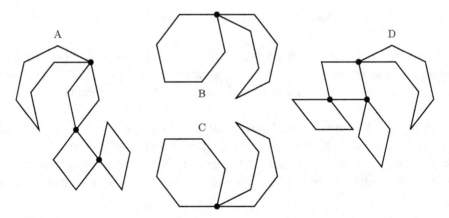

16.6: Hinged pieces for four heptagons to one

For four octagons to one, there is a simple unhingeable 8-piece dissection: Cut a 45°-rhombus out of each small octagon and rearrange the pieces. Accomplishing the same result with hinges seems to take more work. I cut a pair of the small octagons and hinge them in the same way, and I do the same for the remaining pair. My 12-piece hinge-snug dissection claims the pot in Figures 16.7 and 16.8.

The odds seem to be stacked against our finding nice four-into-one dissections of 5-pointed and 7-pointed stars, but we can cash in our chips on 8-pointed stars.

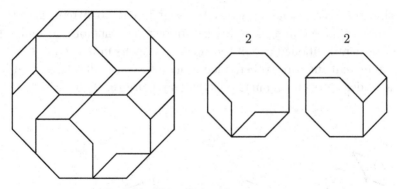

2   2

16.7: Hingeable dissection of four octagons to one

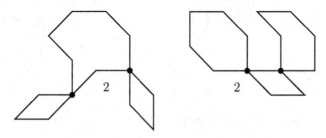

2   2

16.8: Hinged pieces for four octagons to one

Rummaging around in my files from a quarter-century ago, I found a 12-piece unhingeable dissection of four {8/2}s to one. With only four more pieces my 16-piece hingeable dissection wins big in Figure 16.9. I cut all four small stars identically, giving the large star 4-fold rotational symmetry. This dissection is hinge-snug.

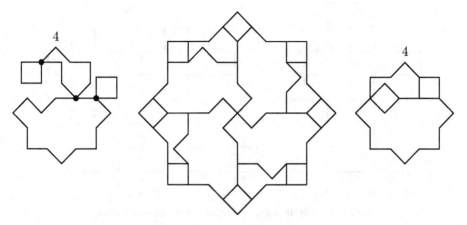

4   4

16.9: Hinged dissection of four {8/2}s to one

157

Finally we deal with four enneagons. Harry Lindgren (1964b) reproduced C. Dudley Langford's simple unhingeable 10-piece dissection. Langford, a London mathematics teacher, positioned one of the small enneagons in the center of the large enneagon and cut the three others identically into pieces that fit around it. The same basic idea works for a hinged dissection (Figure 16.10), for which we use 16 pieces.

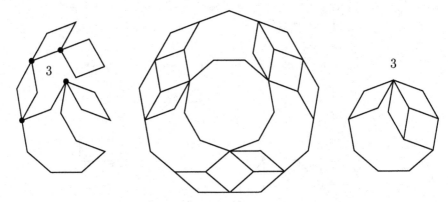

16.10: Hinged dissection of four enneagons to one

If we review these hingeable dissections of four polygons to one, an intriguing pattern emerges: Except for the case when $p = 6$, polygon $\{p\}$ uses $4 * \lceil p/2 \rceil - 4$ pieces. Can a reader find other exceptions to this pattern?

Before we fold, let's try our hand with four Latin Crosses. It is not difficult to find an 8-piece unhingeable dissection. Of course, a hinged dissection is more

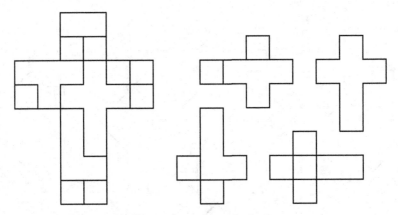

16.11: Hinged dissection of four Latin Crosses to one

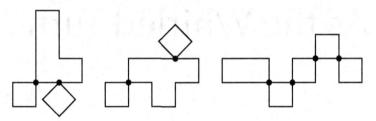

16.12: Hinged pieces for four Latin Crosses to one

challenging. The best I have found is the 12-piece dissection that I turn face up in Figures 16.11 and 16.12.

Puzzle 16.1 Find a 12-piece hinged dissection of four Greek Crosses to one.

CHAPTER 17

As the Whirled Turns

For certain privileged classes of polygons and stars, their rich internal structure leads to *favorable* dissections of n figures to one, for appropriate values of n. Figures containing squares in their structure qualify for favorable two-into-one dissections, since the diagonal of a square is $\sqrt{2}$ times the side. Figures containing equilateral triangles qualify for three-into-one, since the altitude of the triangle is $\sqrt{3}/2$. Other values of n have classes too.

Now, however, we plan to hinge and swing dissections. Can these classes survive the almost irresistible (centrifugal) forces that we shall bring to bear? Can our $\{p\}$s and $\{p/q\}$s find true hingeability under difficult circumstances? Will these figures continue to maintain their symmetry and beauty while spinning out of control? Tune in to catch each installment of "As the Whirled Turns." — *Warning: This program contains polygons more complicated than triangles, squares, and hexagons. Viewer discretion is advised.*

The action starts with the class of figures that contain squares in their internal structure. The most prominent of this two-into-one class is the dissection of two octagons to one. There are 8-piece unhingeable dissections by C. Dudley Langford (1967b) and Ernest Irving Freese, as claimed by Langford (1967a) and given by Lindgren (1964b). Early in my investigation of hinged dissections I had found a 12-piece hingeable dissection based on Freese's dissection and possessing nice symmetry.

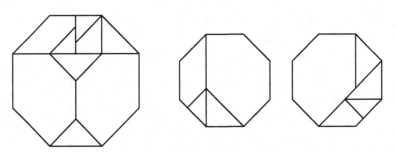

17.1: Theobald's two octagons to one

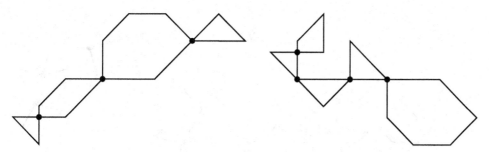

17.2: Hinges for two octagons to one

Then Gavin Theobald made a shocking discovery. He broke both symmetry and rhombuses in producing the marvelous 9-piece hingeable dissection of Figure 17.1. Not only did Gavin split squares to get the $\sqrt{2}$, he also busted up two of the 45°-rhombuses, to give more flexibility in hinging. He hinged the pieces so that they would neatly fold around and fill the area at the top of the large octagon. What a remarkable achievement! Cover up the hinged pieces in Figure 17.2, and see if you can divine how to hinge the pieces.

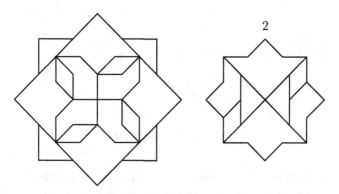

17.3: Two {8/2}s to one

Can two {8/2}s find greater symmetry together than did their cousins, the octagons? Harry Lindgren (1964b) gave a clever 11-piece unhingeable dissection. My 16-piece hinged dissection in Figures 17.3 and 17.4 has 2-fold replication and 2-fold rotational symmetry in the small stars and 4-fold rotational symmetry in the large star.

Puzzle 17.1 There is an attractive hinged dissection of two {8/2}s to one in which only one of the two small {8/2}s is cut into pieces and the pieces are cyclicly hinged. Find it.

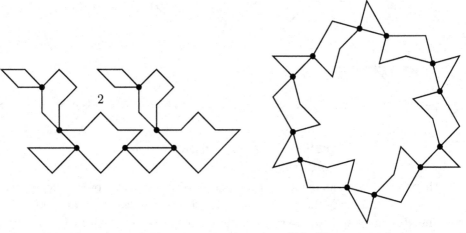

17.4: Hinges: two {8/2}s to one

17.5: Hinges: two dodecagons

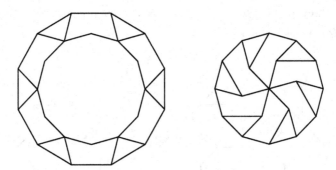

17.6: Two dodecagons to one

Let's next turn to two dodecagons to one. According to Langford (1967b), Ernest Irving Freese described a 12-piece unhingeable dissection. Trading 6-fold rotational symmetry for 3-fold, Lindgren (1964b) modified a 13-piece unhingeable dissection by mathematics teacher Joseph Rosenbaum (1947) to get a 10-piece unhingeable dissection. Will hinging ruin this remaining symmetry, or can we regain the previously squandered beauty? My 13-piece hinged dissection in Figures 17.6 and 17.5 uses the dodecagonal structure that Rosenbaum took advantage of and preserves its symmetrical heritage. The cyclic hinging is hinge-snug. Sharp-eyed readers will recognize that this dissection uses the same approach as the large family of dissections in Chapter 9. In fact, the discovery of the dissection here helped to lead to my discovery of the pattern in that chapter.

Can hingeability withstand the pressures in the 12-pointed domain? I was evidently the first to deal with dissections of two {12/2}s to one, giving a 13-piece

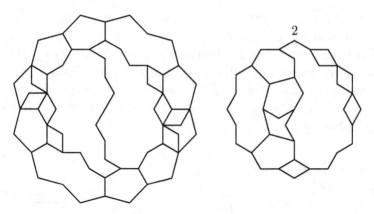

17.7: Hingeable two {12/2}s to one

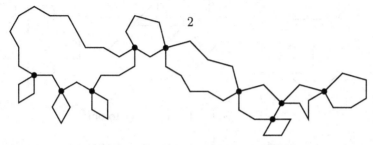

17.8: Hinges for two {12/2}s to one

unhingeable dissection (see Frederickson 1972b). In that dissection I left one of the small {12/2}s uncut and cut the other into pieces that I then arranged around the uncut star. I rely on similar cuts in my 20-piece hinged dissection, shown in Figures 17.7 and 17.8. To migrate rhombi from regions of overlap to holes that they

17.9: Hinged two {12/4}s to one

can fill, I use half of each small {12/2} to fill in the interior of the large {12/2} and the other half to form the exterior.

The patriarch of the two-into-one class is the dissection of two {12/4}s to one. I gave a 19-piece dissection in (1972b) and then an 18-piece dissection in (1972a). The 19-piece dissection is hingeable, and I include its hinged pieces in Figure 17.9. This dissection is surprising in that it takes only one more piece than the current best unhingeable dissection.

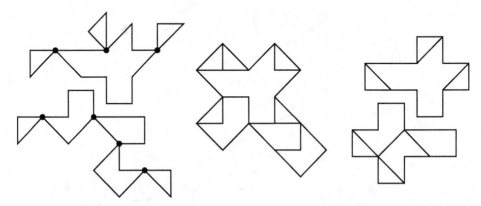

17.10: Hinged two Latin Crosses to one

Hingeable or unhingeable, dissections of two Latin Crosses to one are apparently unprecedented. Since a Latin Cross consists of six squares, it can lean on the dissection of two squares to one. My 9-piece hingeable dissection applies this idea in Figure 17.10. An interesting feature is that one of the hinges can be frozen. The frozen hinge is in the lower small Latin Cross in Figure 17.10, between the top piece and the piece to its right.

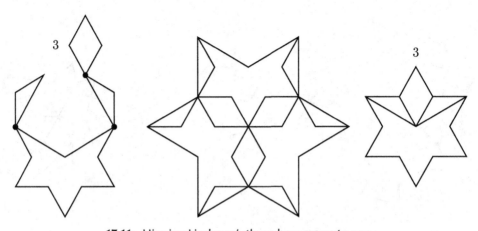

17.11: Hinging Lindgren's three hexagrams to one

If squares foster sedate pairings of figures, equilateral triangles encourage ménages à trois. Our three-to-one dissections start with hexagrams. Lindgren (1964b) gave a 13-piece unhingeable dissection, but he soon found a 12-piece dissection (1964a), which Martin Gardner included in his book (1969). Lindgren's 12-piece dissection (on the right in Figure 17.11) is hingeable, although Lindgren gave no indication of this. Each of the hinged assemblages is linearly hinged.

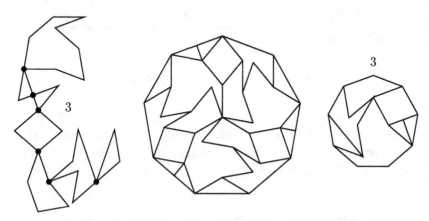

17.12: Hinged dissection of three enneagons to one

Three enneagons to one is difficult to juggle. H. Martyn Cundy and C. Dudley Langford (1960) gave a 21-piece unhingeable dissection, improved by Harry Lindgren (1964a) to 18 pieces. In (1972c) I described a 15-piece dissection, but Robert Reid and Anton Hanegraaf independently found 14-piece dissections. I adapt many of these tricks in finding a 21-piece hinged dissection (Figure 17.12). It has 3-fold replication symmetry in the small enneagons and 3-fold rotational symmetry in the large one. Although wobbly hinged, each enneagon is linearly hinged and hinge-snug.

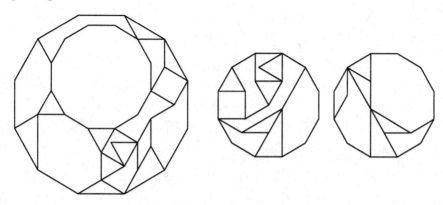

17.13: Three dodecagons to one

165

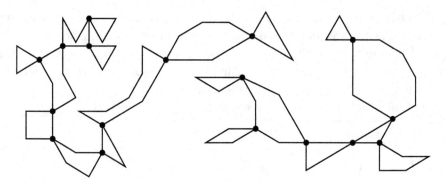

17.14: Hinges for three dodecagons to one

For three dodecagons to one, Lindgren (1964b) gave a 15-piece unhingeable dissection. Robert Reid and Anton Hanegraaf independently discovered 14-piece unhingeable dissections. I found a 23-piece hinged dissection, but this has been superseded by Anton's 20-piece wobbly hinged dissection in Figures 17.13 and 17.14.

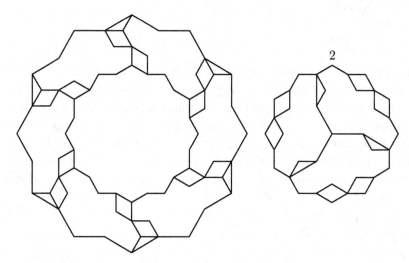

17.15: Three {12/2}s to one

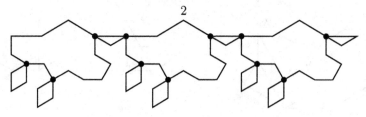

17.16: Hinges for three {12/2}s to one

He based his dissection on his 14-piece unhingeable dissection, cutting two of the dodecagons to fit around the third. Could any solution be more devious?

In (1972b) I gave an 18-piece unhingeable dissection of three {12/2}s to one. In Figures 17.15 and 17.16, I give a 25-piece hinged dissection. I leave one {12/2} uncut, and cut each of the other two identically into 12 pieces. Two small rhombuses hang off of each large piece. One rhombus swings out of the way of the uncut {12/2} and into the former position of the other rhombus, which swings out further.

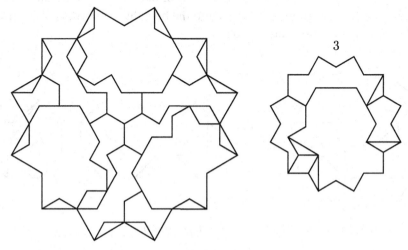

17.17: Hanegraaf's three {12/3}s to one

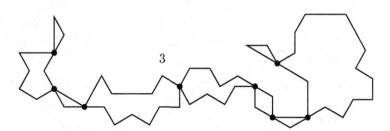

17.18: Hinges for three {12/3}s to one

The next dissection will leave us spinning. In (1972b) I gave a 24-piece unhingeable dissection of three {12/3}s to one. Using a different approach, I improved the dissection by 6 pieces in (1974). Then I found a 36-piece hingeable dissection that is based on an approach similar to my original unhingeable dissection. I wondered if a reader could improve on the 36 pieces. Prompted by this challenge, Anton Hanegraaf found a 27-piece hinged dissection that is loosely based on my (1974) dissection. His linearly hinged dissection, which amazes us in Figures 17.17 and 17.18, has a beautiful 3-fold rotational symmetry.

167

While triangle-based polygons have been engaged in their ménages, our square-based class has found itself similarly enmeshed. There is a favorable dissection of three into one for those polygons and stars whose number of vertices is a multiple of 8. Let's rejoin them to see how the octagon fares. In (1974), I gave a 10-piece un-hingeable dissection of three octagons to one. My 14-piece hinged dissection bares itself in Figures 17.19 and Figure 17.20. I cut two of the small octagons into four pieces each that easily fit into the large octagon. The third octagon is a bit tougher. Casting aside all scruples, I cut a rhombus out of one of the pieces that has a side length of $\sqrt{3}$. Another rhombus swings into the hole thus created, and the original rhombus fills a hole in the large octagon.

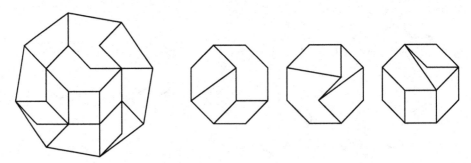

17.19: Three octagons to one

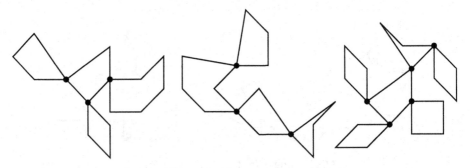

17.20: Hinges for three octagons to one

So again an octagon dissection has exhibited no sense of symmetry. Yet, the {8/2} remains symmetrically stoic, as we now see. In (1974), I gave a 16-piece un-hingeable dissection of three {8/2}s to one. Then Robert Reid one-upped me with a 15-piece unhingeable dissection. I found a 25-piece hinged dissection, which Anton Hanegraaf modified slightly to the 23-piece dissection that sits imperiously in Figures 17.21 and 17.22. We cut and hinge two of the small stars in the same way,

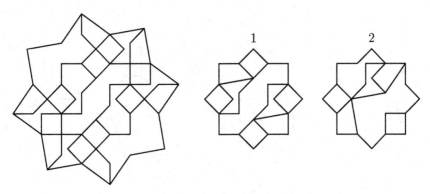

17.21: Three {8/2}s to one

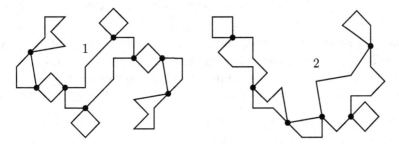

17.22: Hinges for three {8/2}s to one

with each contributing half of the eight sides of the large star. We then cut the third small star with 2-fold rotational symmetry. The third small star goes in the center of the large star, with a portion cut out of two parts of the outside to accommodate three pieces from the outside of each of the other two identically cut stars. We cut the portion in such a way as to finish the outline of the large star.

The square-based class is ripe for other subplots, such as five-to-one dissections. Byzantine is not the word to describe them, because they hark back to Abū'l-Wafā's dissection of five squares to one. Again, we first encounter the octagons. I found a 17-piece unhingeable dissection in which I cut four octagons identically and left the fifth uncut. The 25-piece hinged dissection in Figure 17.23 follows a similar approach, but two more pieces are in each of the four cut octagons.

The rhombic piece on the far right swaps around as we did in the dissection of three octagons to one. Although the rhombic piece on the left cannot swing into a hole that is left for it, it can swing into a hole left when the piece on the right swings out of the small octagon. Conveniently, the piece on the right then fills the hole that the piece on the left could not. Has rhombus-swapping ever been more blatant?

169

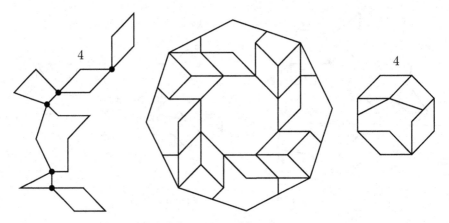

17.23: Hinged dissection of five octagons to one

Can we bring five {8/2}s to one under a similar control? I found a 21-piece un-hingeable dissection in which I did not cut one {8/2} and cut the other four identi-cally. Anton Hanegraaf has modified that dissection to give the 29-piece hingeable dissection in Figure 17.24. He uses two more pieces in each of the four {8/2}s that are cut.

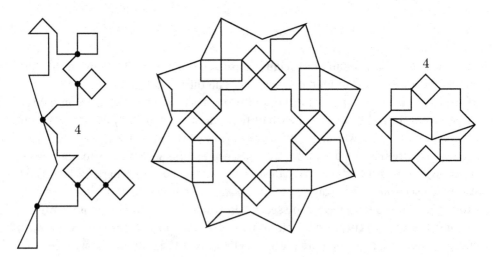

17.24: Hinged dissection of five {8/2}s to one

The trick is in shifting around small squares from places where they are in the way to places where they are needed. Anton's approach seems to be the only eco-nomical way to accomplish this. Unfortunately, the pieces are wobbly hinged.

Will we find five dodecagons to one to be as even-tempered? I found a 21-piece unhingeable dissection in which I did not cut one of the dodecagons and cut the other four identically. Anton Hanegraaf adapted my basic approach to produce the 29-piece hingeable dissection in Figure 17.25. Once again, the dissection is wobbly hinged. Perhaps a reader can find a 29-piece dissection that does not need the wobbly hinges.

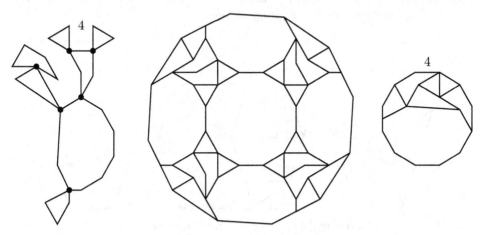

17.25: Hinged dissection of five dodecagons to one

To escape the influence of these worldly polygons, we return to Latin Crosses. In (1977), Cornell University physicist Veit Elser and Michael Goldberg showed how to cover a cube with five Latin Crosses. Their covering converts to a 15-piece unhingeable dissection of five Latin Crosses to one. I can do better, and with hinges, too! Trial and error led to my 14-piece hing-snug dissection in Figure 17.26.

It's now time for a major turn in the plot, with the introduction of a new class. If a polygon or star has a number of vertices that is divisible by 5, then we have left

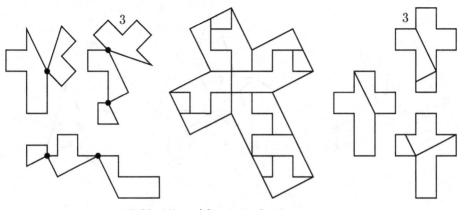

17.26: Hinged five Latin Crosses to one

the door open for a favorable five-to-one dissection. First, pentagons make their entrance. Langford (1956) gave an elegant 12-piece unhingeable dissection of five pentagons to one. A 15-piece dissection by Lindgren (1964b) is hingeable, although he never pointed that out. Yet another hinged dissection, this one hinge-snug, appears in Figure 17.27. The large pentagon has 5-fold rotational symmetry and the small pentagons have 5-fold replication symmetry.

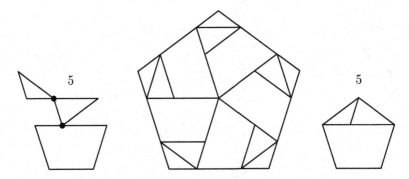

17.27: Hinged dissection of five pentagons to one

Puzzle 17.2 Hinge the same set of pieces in Figure 17.27 differently so that they still form a large pentagon or five small ones.

Next upon the scene are the pentagon's cousins: pentagrams. Lindgren (1964b) attributed a symmetrical 20-piece unhingeable dissection of five pentagrams to one to Ernest Irving Freese. In (1974) I reduced the number of pieces by two, but at the loss of symmetry. I have since found a hinged dissection (Figure 17.28) that

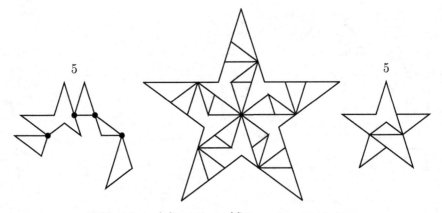

17.28: Hinged dissection of five pentagrams to one

restores the symmetry and uses 25 pieces. The key to the dissection is determining the shape of the nontriangular piece and its position within the large pentagram. Once I made a felicitous choice, the rest of the pieces fell easily into place.

The third illustrious member of this class is the decagon. In (1974) I gave a 17-piece unhingeable dissection of five decagons to one. The key idea was to find a line segment of length $\sqrt{5}$ in a small decagon and thus have each small decagon contribute two of the sides of the large decagon. The remaining pieces of the small decagons then fill in the interior of the large decagon. The 25-piece wobbly hinged dissection in Figure 17.29 follows a similar approach. As shown in the middle and the right, I cut each small decagon into five pieces.

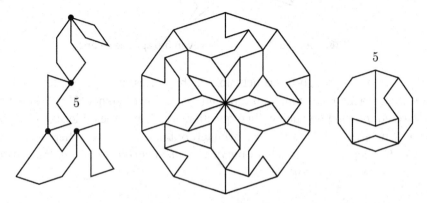

17.29: Hinged dissection of five decagons to one

Meanwhile, our class of triangle-based figures is back in action, in ever greater numbers! Since hexagrams consist of triangles, we expect an advantageous hinge-able dissection of seven to one. Mine is based on the 19-piece unhingeable dissec-tion by Alfred Varsady. I did not cut one of the small hexagrams and cut the six others identically. My 25-piece hinge-snug dissection in Figure 17.30 has one addi-tional piece in those six hexagrams, allowing the rhombuses to be folded around to fill in holes in the large hexagram.

In a final turn of the plot, a dissection of eight heptagons to one swings be-fore our eyes. I found a 22-piece unhingeable dissection, which was a modification of Alfred Varsady's 29-piece dissection. Both of these dissections have one hep-tagon uncut and the other seven identically cut. The 36-piece hinged dissection that catches our eye in Figure 17.31 follows a similar approach, except that there are two more pieces in each of the seven cut heptagons than in my unhingeable dis-section. However, I turn over no pieces, in contrast to the seven that I turned over in my unhingeable dissection. I use the rhombic piece to help swap pieces: the trape-zoidal piece swings around to fit into the hole from which the rhombic piece is cut.

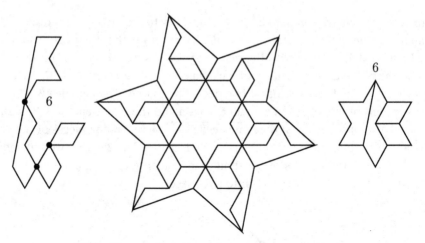

17.30: Hinged dissection of seven hexagrams to one

We have now come to the conclusion of this series of episodes, in which our figures have reached a turning point in their geometric existence. Much remains to be answered, hinging on such crucial questions as: Can three dodecagons achieve greater symmetry? Why have {8/3}s been left out in the cold? Can heptagons both swing *and* be hep? — *Stay tuned for the further dizzying adventures of the polygons and stars on "As the Whirled Turns."*

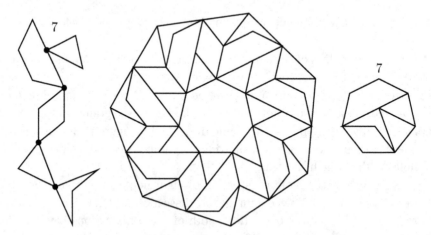

17.31: Hinged dissection of eight heptagons to one

Curious Case, part 5: A Quorum of Quibbles

Our curious case turns in part on whether Henry Dudeney used careful wording to make it appear that he had discovered the 4-piece solution. Because he had written a weekly puzzle column for six years, it should be no surprise that he had developed a sensitivity to careful wording and the creative uses of quibbles. So let's examine his discussion of the triangle and square puzzle and study the wording in a number of his early puzzles.

First, his columns of April 1902, in which he posed the puzzle and announced the 4-piece solution:

> "Several competitors correctly judged that if five were the smallest possible number of pieces into which an equilateral triangle may be cut to form a square I should not have written of the 'inherent difficulty' of the problem."
>> The use of the word "should" is ambiguous; it can express either obligation or what is probable. A puzzle columnist might sometimes write what he should not write, namely that he considers a problem solution difficult (i.e., the 5-piece solution) when others might consider it easy.

> "It seems that I was right in believing that it would be solved, though only one competitor succeeded in getting at 'the heart of the mystery.'"
>> Dudeney did not say the less ambiguous "though only one competitor succeeded in discovering the 4-piece solution that I had in mind, and thus getting at 'the heart of the mystery.'"

> "I have certainly had the problem in my notebooks for some years, but have kept it back principally on account of its inherent difficulty."
>> Having had the problem in his notebooks does not necessarily mean that he had the 4-piece solution in those notebooks.

Similarly, the wording of the Haberdasher's Puzzle in the "Canterbury Puzzles" in November 1903 leaves an opening:

> "we have found that the feat may really be performed in so few as four pieces"
>> Does this mean he discovered the 4-piece solution, or just learned of its existence (from McElroy)? Dudeney could have stated more directly that he had "discovered" a 4-piece solution.

Care in the choice of words was a significant trait of Dudeney's. He used it effectively in posing a number of puzzles. The following are examples from his column in the *Weekly Dispatch* in which careful interpretations of wording are crucial to the puzzle solution.

1. In "The Map Puzzle," on September 6, 1896, Dudeney asked essentially how many colors it would take to color a given map. His solution two weeks later relied on the trick of mixing two colors to get a third.

2. In his puzzle, "The Eleven Bears," on August 6, 1899, Dudeney asked how eleven bears could be arranged so that there were as many rows with four bears in a row as possible. In the solution two weeks later, he revealed that a twelfth bear whose shooting had been described earlier in the story was also to have been used in the solution.

3. In his puzzle, "The Eight Bridges," on March 24, 1901, Dudeney presented a variation on the Königsberg bridge problem. But he stated it in a way that left open the possibility of walking to the head of a river, around its source, and back!

4. In the puzzle of "The Gardener and the Cook," on April 1, 1900, Dudeney asked who had won a 100-foot race between the gardener and the cook, assuming that the gardener ran three feet in each bound while the cook only two feet but that she made three bounds to his two. In the solution four weeks later, Dudeney anticipated modern-day gender correctness by pointing out that the gardener could be a woman and the cook a man!

Dudeney was capable of pushing modesty aside and clearly claiming a result for himself. In the solution to "Magic Squares of Two Degrees," on March 29, 1903, he stated unequivocally that he had discovered the solution: "The first diagram below is my own arrangement; the second is Mr. Baxter's." And later: "The difficulty lies entirely in discovering for yourself (as Mr. Baxter and I had each of us to do) the laws that govern it."

A final example of Dudeney's awareness of careful wording is his solution to "Those Fifteen Sheep" on July 26, 1903:

> Now, one thing is very clear from the start – that the answer, whatever it may be, is a quibble of some sort, and our competitors generally have gone to work with this idea. It is evident that if there is to be found any solution at all, it must lie in the existence of some flaw in the wording. I have suggested to my readers over and over again that it is often of the first importance in solving puzzles to be quite sure we have read the conditions accurately.... Here are the exact words: "Place fifteen sheep in four pens so that there shall be the same number of sheep in each pen." We were not told that the pens were necessarily empty. In fact, if the reader will refer back to the illustration he will see that one sheep is already in one of the pens. It is just at this point that the wily farmer said to me: "*Now* I am going to start placing the fifteen sheep."

So Dudeney was alert to the possibilities of the quibble and was practiced in its use as well. *Now* how should we read his columns of April 1902?

CHAPTER 18

A New Breed
of Swingers

It was my privilege to have come to know three dissection experts – Stuart Elliott, Alfred Varsady, and Robert Reid – who all exploited new geometric relationships in their dissections. I called them, and their dissections, the "new breed." Almost all of those dissections are unhingeable, so we now have new opportunities open to us. With Stuart deceased and with Alfred and Robert less active, I will certainly not call my three old friends a "new breed of swingers." But in their honor, I will so call the hinged dissections based on those relationships.

Elliott (1982) saw that the special relationships that lead to auspicious dissections can be combined with the favorable dissections of *n* identical polygons to one, yielding what I called *propitious* dissections. Let's boogie with a few of these here.

Matching up the relationship between a dodecagon and a square with three dodecagons to one gives a dodecagon to three squares. The corresponding dissection illustrates a rough approximation of 3 for π. Such a dissection proof appears in an illustration prepared for Palace edition of the *Chiu Chang Suan Shu* (*Nine Chapters on the Mathematical Art*) by the eighteenth-century Chinese scholar Tai Chen: see (Needham 1959) and (Shen et al. 1999). More recently, the Hungarian mathematician Josef Kürschák (1899) gave a similar proof, and Elliot (1985) gave an unhingeable 9-piece dissection. If we use wobbly hinges, then a 9-piece hinged dissection is possible (see Figures 18.1 and 18.2). Swing several pieces to mate with the

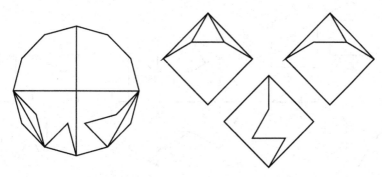

18.1: One dodecagon to three squares

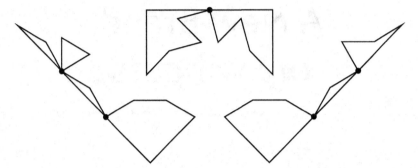

18.2: Wobbly hinges for one dodecagon to three squares

dodecagonal quarter in the upper left, using the smaller pieces from the bottom of the dodecagon. Do the same thing with the dodecagonal quarter in the upper right. What is left are two pieces that mate together to give the third square. With trial and error we can find a 10-piece hingeable dissection that is not wobbly.

Puzzle 18.1 Find a 10-piece hingeable dissection of a dodecagon to three squares that is not wobbly hinged.

Interbreeding the previous relationship with two squares to one leads to a dissection of a dodecagon to six squares. Robert Reid found a 14-piece unhingeable dissection, which I modify to give the 16-piece hinged dissection that stands symmetrically in Figure 18.3. Robert found the four squares that contain two pieces each, but he produced each of the remaining two squares from three unhingeable pieces. With one piece more per square I have made these last two squares hingeable. This dissection is hinge-snug.

Mating the relationship between a hexagram and a triangle with three hexagrams to one, Stuart Elliott (1982) found a 9-piece unhingeable dissection of three

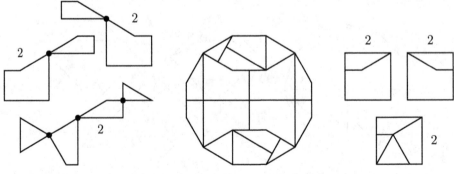

18.3: Hinged one dodecagon to six squares

hexagrams to a triangle. With 3-fold replication symmetry for the hexagrams and 3-fold rotational symmetry in the triangle, it is tempting to believe that the simple approach is the best. However, Robert Reid (1987) discovered a clever 8-piece unhingeable dissection that destroyed the symmetry as it saved one piece. So when we look for a hingeable dissection, we can find a relatively simple 15-piece dissection that contains the same replication and rotational symmetry as Elliott's. However, now we know to be cautious. In Figures 18.4 and 18.5, I give a 13-piece hinged dissection that wrecks the symmetry for hingeable dissections, much as Reid's did for unhingeable dissections.

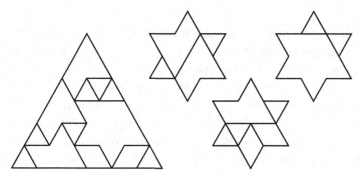

18.4: Three hexagrams to one triangle

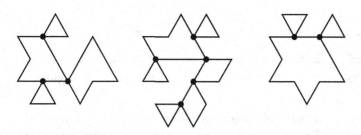

18.5: Hinges for three hexagrams to one triangle

Selecting the relationship between a {12/2} and a hexagon to pair with three hexagons to one, Elliott (1982) found a 13-piece unhingeable dissection of one {12/2} to three hexagons. Then Reid applied my (1972d) approach to give an 11-piece unhingeable dissection. After finding the 15-piece hinged dissection in Figure 18.6, I realized that superposing tessellations produces this dissection. Do you see it?

Puzzle 18.2 Find the superposition of tessellations that leads to the dissection of a {12/2} to three hexagons.

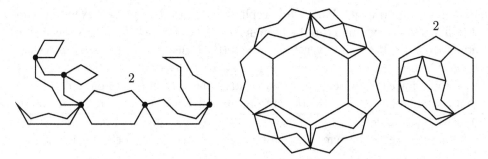

18.6: Hinged dissection of one {12/2} to three hexagons

Identifying different types of special relationships, Alfred Varsady (1986) introduced an altogether new breed. He investigated how to build 5-pointed and 10-pointed figures out of the two types of isosceles triangles in Figure 18.7. The first triangle has two angles of $\pi/5$ and the second has two angles of $2\pi/5$. I have called these the *iso-penta triangles,* $\{3_{1/5}\}$ and $\{3_{2/5}\}$, respectively. Varsady used a simple expansion to give versions of $\{3_{1/5}\}$ and $\{3_{2/5}\}$ that have sides longer by a factor of ϕ (the golden ratio). Figure 18.8 shows that two $\{3_{1/5}\}$s plus one $\{3_{2/5}\}$ form a larger $\{3_{1/5}\}$ and that one each of $\{3_{1/5}\}$ and $\{3_{2/5}\}$ similarly form a larger version of $\{3_{2/5}\}$.

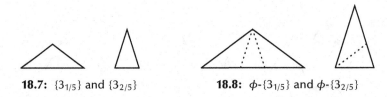

18.7: $\{3_{1/5}\}$ and $\{3_{2/5}\}$ **18.8:** ϕ-$\{3_{1/5}\}$ and ϕ-$\{3_{2/5}\}$

The *signature* (a, b) of a figure gives the number a of $\{3_{1/5}\}$s followed by the number b of $\{3_{2/5}\}$s. For example, the signature of a 1-pentagon is $(3, 1)$ and the signature of a 1-pentagram is $(2, 4)$. When each figure in a dissection has a signature consisting of integers, the dissection seems to need fewer pieces. I have termed such dissections *advantageous.*

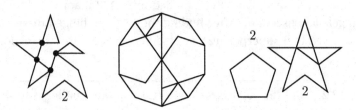

18.9: Hinged two pentagons and two pentagrams to one decagon

180

Alfred Varsady investigated various combinations of identical pentagons and identical pentagrams into a large decagon. I will now explore one such bloodline. Varsady found an unhingeable 6-piece dissection of two 1-pentagons and two 1-pentagrams to a 1-decagon, whose signature is (10, 10). I have found a 12-piece hinged dissection (Figure 18.9).

Varsady also found an 18-piece unhingeable dissection of eight pentagons and three pentagrams to a ϕ-decagon, whose signature is (30, 20). I found a 19-piece hinged dissection (Figures 18.10 and 18.11).

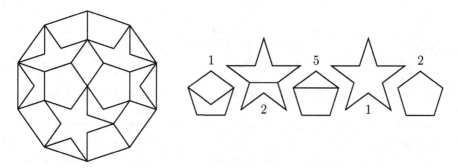

18.10: Eight pentagons and three pentagrams to one decagon

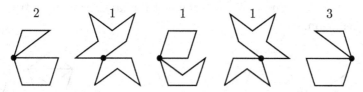

18.11: Hinges for eight pentagons and three pentagrams to one decagon

Varsady further found an 36-piece unhingeable dissection of 22 pentagons and seven pentagrams to a ϕ^2-decagon, whose signature is (80, 50). Since five sides of Varsady's decagon result from pentagrams, which we cannot similarly cut in a hingeable dissection, my 42-piece hinged dissection in Figure 18.12 seems relatively good.

In 1993, Alfred Varsady propagated two types of pentagram dissections similar to the decagon dissections that we have already seen. One bloodline extends in a convenient way to hingeable dissections, as I shall now demonstrate. The first in Alfred's family is an 8-piece unhingeable dissection of two 1-pentagons and a 1-pentagram to a ϕ-pentagram, with signature (8, 6). I have found the 10-piece hingeable dissection that winks at us in Figure 18.13. Varsady cut his 1-pentagram in the same way, but his dissection of the two pentagons was not hingeable.

181

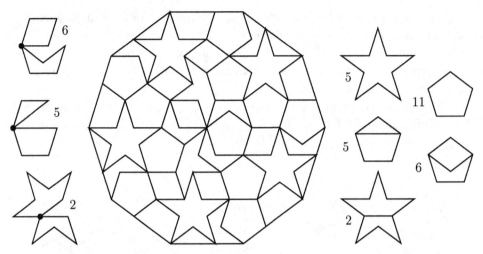

18.12: Twenty-two pentagons and seven pentagrams to one decagon

The next in this bloodline is Varsady's 14-piece unhingeable dissection of six 1-pentagons and two 1-pentagrams to a ϕ^2-pentagram, with a signature of $(22, 14)$. My 21-piece hinged dissection skirts with symmetry in Figures 18.14 and 18.15. The difficult part of hinging the dissection is creating the points of the large pentagram. I have accomplished this by moving the small pentagrams to the interior of the large one and creating most of a point of the large pentagram from a corresponding pentagon.

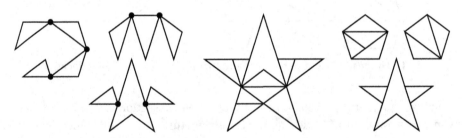

18.13: Hinged two pentagons and a pentagram to a ϕ-pentagram

The third pentagram dissection in the bloodline is a 31-piece unhingeable dissection of sixteen 1-pentagons and five 1-pentagrams to a ϕ^3-pentagram, whose signature is $(58, 36)$. I have found the 41-piece hinged dissection in Figure 18.16. Given Varsady's unhingeable dissection and the hingeable dissection in Figure 18.14, a hingeable dissection is not so hard. We simply place the pentagrams at the bases of the points, as in Varsady's dissection, and fill in the extremities of the points as in our previous dissection.

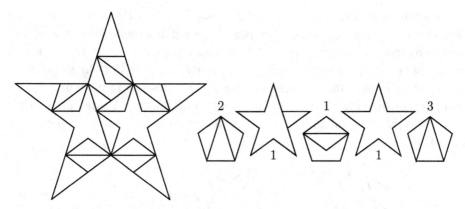

18.14: Six pentagons and two pentagrams to a ϕ^2-pentagram

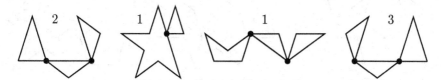

18.15: Hinges for six pentagons and two pentagrams to a ϕ^2-pentagram

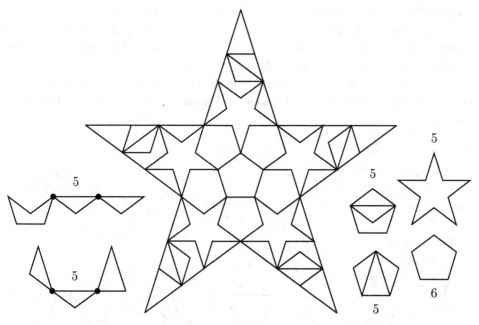

18.16: Hinged sixteen pentagons and five pentagrams to a ϕ^3-pentagram

In addition to these families, Alfred Varsady produced a wealth of new species. One dissection that he sent me is a 15-piece unhingeable dissection of four pentagons and four pentagrams to a {10/3}, whose signature is (20, 20). Even without 20-20 eyesight, my 16-piece hinged dissection in Figure 18.17 looks great. An interesting feature of this dissection is that it simplifies to a 14-piece unhingeable dissection by merging each 36°-rhombus with a neighboring 5-sided piece. Thus, finding a good hingeable dissection led to finding an improvement in an unhingeable dissection.

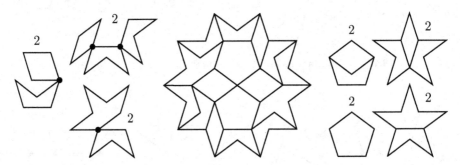

18.17: Hinged four pentagons and four pentagrams to a {10/3}

After seeing Alfred Varsady's creations, Robert Reid sent me a 23-piece unhingeable dissection of thirteen pentagons and four pentagrams to a ϕ^3-pentagon, whose signature is (47, 29). I have found the 30-piece hinged dissection that stands majestically in Figures 18.18 and 18.19. This dissection has reflection symmetry, as long as we do not consider the position of the hinges.

Robert also gave a 16-piece unhingeable dissection of eight pentagons and two pentagrams to a large pentagon. It takes at most one more piece to give a hingeable dissection.

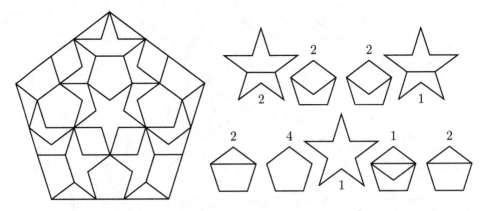

18.18: Thirteen pentagons and four pentagrams to a ϕ^3-pentagon

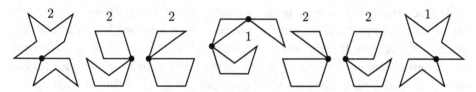

18.19: Hinges for thirteen pentagons and four pentagrams to a ϕ^3-pentagon

---

Puzzle 18.3 Find a 17-piece hingeable dissection of eight pentagons and two pentagrams to a large pentagon.

Alfred Varsady also sent me a 54-piece unhingeable dissection of 27 pentagons and eight pentagrams to a large pentagon; the signature of the large pentagon is $(97, 59)$. I have found the 60-piece hinged dissection that balances so delicately in Figure 18.20. This dissection converts rather easily into a 53-piece unhingeable dissection, a saving of one piece over Varsady's dissection. Saving six of the pieces is really easy, since there are six places where a pentagon can replace a rhombus and a matching concave piece. Can the reader find the remaining saving of one piece? Try replacing a rhombus and three concave pieces with a pentagon and two concave pieces.

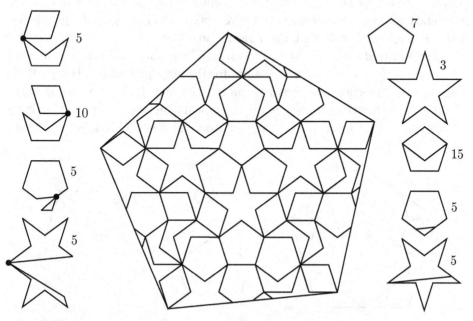

18.20: Twenty-seven pentagons and eight pentagrams to a large pentagon

185

Varsady found a dissection of five pentagrams and two pentagons to a {10/3} (Figure 18.21). Here the pentagons do not have the same side length as the pentagrams and the {10/3}. Split one pentagon into five equal isosceles triangles and then arrange them around the other pentagon. Split the five pentagrams into two pieces each and arrange them around on the periphery. The result is hingeable.

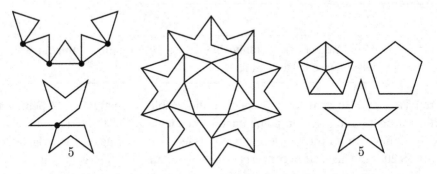

18.21: Two pentagons and five pentagrams to one {10/3}

Varsady (1986, 1989) gave 7-piece unhingeable dissections of pentagons realizing $1^2 + \phi^2 + (\phi^2)^2 = (2\phi)^2$. The signatures for the pentagons are $(3, 1)$, $(7, 4)$, $(18, 11)$, and $(28, 16)$, respectively. I found an 8-piece almost-hingeable dissection, starting with the ϕ^2-pentagon uncut in the center of the 2ϕ-pentagon. But the last piece ends up turned the wrong way. I have since found the 9-piece hinged dissection of Figures 18.22 and 18.23. Can a reader do better?

In 1998, Alfred Varsady sent me an 8-piece unhingeable dissection of pentagons for $1^2 + \phi^2 = \left(\sqrt{1 + \phi^2}\right)^2$. The signature for the large pentagon is $(10, 5)$. My 9-piece hinged dissection responds in Figures 18.24 and 18.25. The trick is in the handling of the isosceles triangle in the small pentagon. We would like to swing it into an appropriate space in the large pentagon, but it is oriented the wrong way.

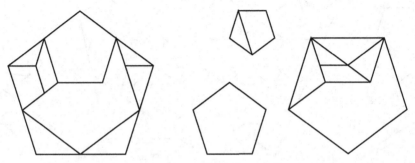

18.22: Pentagons for $1^2 + \phi^2 + (\phi^2)^2 = (2\phi)^2$

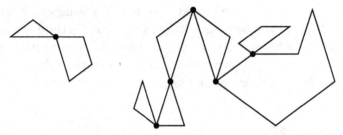

18.23: Hinges for pentagons for $1^2 + \phi^2 + (\phi^2)^2 = (2\phi)^2$

Cut a piece out of the medium-sized pentagon that will swing into the desired space, and swing the triangle from the small pentagon into the hole that is created. This swings the triangle around the small pentagon in the other direction, which is evidently what is needed. We can derive an 8-piece unhingeable dissection from my hingeable one if we do *not* cut the piece out of the medium-sized pentagon.

Also in 1998, Alfred sent me a 12-piece unhingeable dissection of three 1-pentagons and a ϕ-pentagram to a larger pentagon; the signature of the large pentagon is

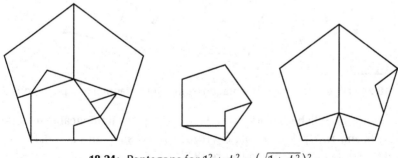

18.24: Pentagons for $1^2 + \phi^2 = (\sqrt{1 + \phi^2})^2$

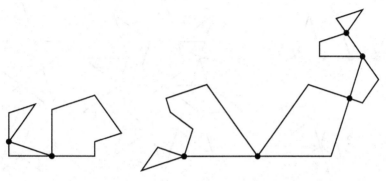

18.25: Hinges for pentagons for $1^2 + \phi^2 = (\sqrt{1 + \phi^2})^2$

187

(17, 9). I have found the 12-piece hinged dissection that swings perilously close to inbreeding in Figures 18.26 and 18.27. Alfred's dissection is close enough to being hingeable that I have been able to modify it. He cut all three 1-pentagons in the same way, but I introduce an extra cut in one to give a piece that swings into an unfilled space. This avoids one cut in the pentagram, so that the number of pieces stays the same.

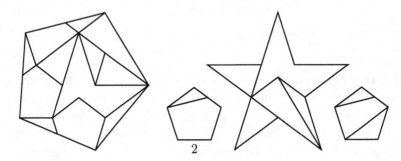

18.26: Three 1-pentagons and a ϕ-pentagram to a pentagon

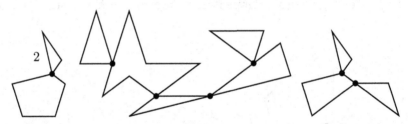

18.27: Hinges for three 1-pentagons and a ϕ-pentagram to a pentagon

In earlier (unpublished) work, Alfred gave a 14-piece unhingeable dissection of two 1-{5/2}s and two ϕ-{5/2}s to a ϕ-{10/4}, whose signature is (24, 16). I found the 20-piece hinged dissection in Figure 18.28.

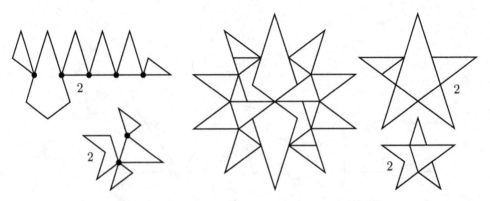

18.28: Two 1-{5/2}s and two ϕ-{5/2}s to a ϕ-{10/4}

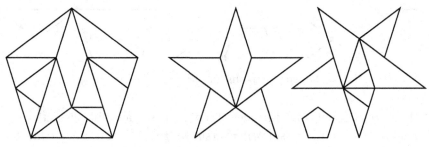

18.29: Two ϕ^2-{5/2}s and a {5} to a ϕ^3-{5}

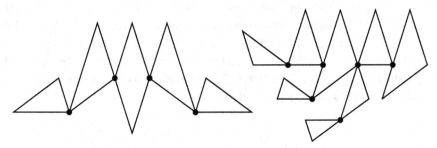

18.30: Hinges for two ϕ^2-{5/2}s and a {5} to a ϕ^3-{5/2}

Producing special relationships in a different way, Swissman Jean Bauer (1999) identified a relationship between two ϕ^2-pentagrams, a 1-pentagon, and a ϕ^3-pentagon. I have found a 10-piece unhingeable dissection for it as well as the 14-piece hinged dissection full of triangles in Figures 18.29 and 18.30.

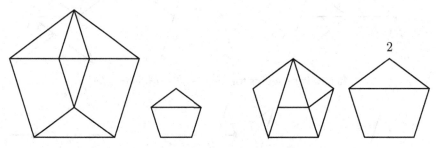

18.31: Pentagons for $(\phi^2)^2 + 1^2 = \phi^2 + \phi^2 + \phi^2$

Varsady (1986) gave a 9-piece dissection for pentagons for $(\phi^2)^2 + 1^2 = \phi^2 + \phi^2 + \phi^2$, shown in Figure 18.31. Surprisingly, the dissection is hingeable, as in Figure 18.32. It is also hinge-snug. Even more interesting, the relation for the dissection is a special case of the following general relation, discovered by Bauer (1999).

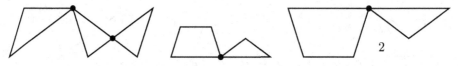

18.32: Hinged pentagons for $(\phi^2)^2 + 1^2 = \phi^2 + \phi^2 + \phi^2$

Let $\{p\}$ be a regular polygon of p sides. Then there is a favorable dissection of $\{p\}$ for $(\sin \pi/p + \sin 3\pi/p)^2 + (\sin \pi/p)^2 = (\sin 3\pi/p)^2 + (\sin 2\pi/p)^2 + (\sin 2\pi/p)^2$.

The last dissection is part of a larger bloodline. The same relationship leads to dissection of pentagrams in place of pentagons. There is a 15-piece unhingeable dissection, and with some work we can find a 22-piece hingeable dissection (Figure 18.33). There are four hinged assemblages, three with labels A, B, C and one that is unlabeled. In the unlabeled set are four pieces that effect a conversion of a triangle to the same triangle but connected differently. This uses Macaulay's dissection, which we have seen in Figure 3.23. The hinged pieces dance in Figure 18.34, with the A, B, and C assemblages labeled.

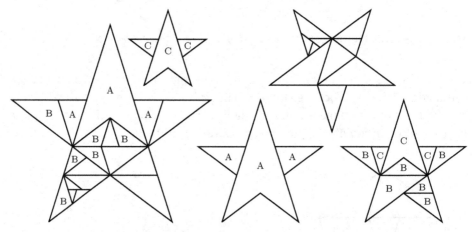

18.33: Pentagrams for $(\phi^2)^2 + 1^2 = \phi^2 + \phi^2 + \phi^2$

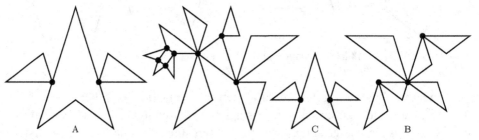

18.34: Hinged pentagrams for $(\phi^2)^2 + 1^2 = \phi^2 + \phi^2 + \phi^2$

I have found a 9-piece hinge-snug dissection of hexagons for $3^2 + 1^2 = (\sqrt{3})^2 + (\sqrt{3})^2 + 2^2$, shown in Figures 18.35 and 18.36. We can view it as a combination of three hexagons to one and hexagons for $1^2 + (\sqrt{3})^2 = 2^2$.

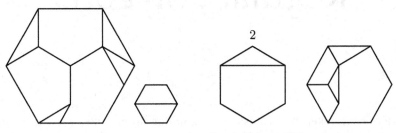

18.35: Hexagons for $3^2 + 1^2 = (\sqrt{3})^2 + (\sqrt{3})^2 + 2^2$

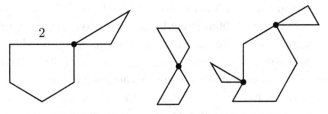

18.36: Hinged hexagons for $3^2 + 1^2 = (\sqrt{3})^2 + (\sqrt{3})^2 + 2^2$

I have not yet found a satisfying hingeable dissection for heptagons – sometimes good breeding takes time. But there is a nice 7-piece hinged dissection for the previous identity applied to triangles. You can even find a hinge-snug version for our puzzle, with which we conclude the chapter.

Puzzle 18.4   Find a 7-piece hingeable dissection of triangles for $3^2 + 1^2 = (\sqrt{3})^2 + (\sqrt{3})^2 + 2^2$.

CHAPTER 19

Not Your Regular Polygons

No readers have taken me to task for promoting an idealized view of geometric beauty. Yet, without a doubt, an average-looking polygon is almost surely not regular. So, let's be politically as well as mathematically correct, turn to irregular polygons, and find within them an inner beauty. Harry Hart (1877) discovered two related dissections of certain similar irregular polygons to a larger copy of the same. One dissection applies when we can circumscribe the polygon, which means that we can draw a circle (the *circumcircle*) so that all vertices of the polygon are on the circle. The second dissection applies when we can inscribe the polygon, so that all sides are tangent to an incircle. Interestingly, the first of Hart's two dissections is hingeable – namely, the one that applies to circumscribed polygons. We designate by $\{\textcircled{p}\}$ a p-sided polygon that we can circumscribe.

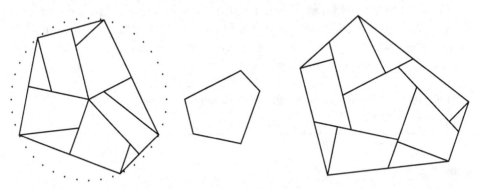

19.1: Hart's hingeable dissection of two circumscribed pentagons to one

We apply Hart's dissection to an irregular pentagon in Figure 19.1. The larger of a pair of such pentagons is on the left, circumscribed by its circumcircle. Hart made a cut from the midpoint of each of the sides to the circumcircle's center, along the side's perpendicular bisector. He then made a cut from each vertex to the next perpendicular bisector in the counterclockwise direction, cutting off a right triangle whose legs are in the ratio of the side lengths of the polygons that he was combining.

19.2: Hinging Hart's two similar circumscribed pentagons to one

The dissection has a nifty cyclic hinging, displayed in Figure 19.2. Hart gave no indication that he was aware that his dissection is hingeable. It is also hinge-snug and grain-preserving. We have already noted in Chapter 9 the connection to an enormous family of dissections. A whimsical view of Figure 19.2 suggests the enclosed area as a body with neck, two arms and two legs. The triangles then become a head, two hands and two feet – in a delicate dance that acknowledges no deficiency in its irregularity.

Now let's get these polygons to swing! We can generalize both of Hart's dissections to handle two similar polygons to two different similar polygons. For polygons

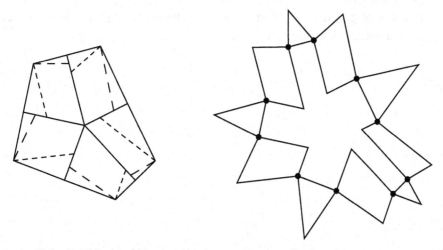

19.3: Deriving and hinging Hart's two circumscribed pentagons to two

that we can circumscribe, the generalization comes about in the following way. We see that each right triangle cut out of the pentagon on the left in Figure 19.1 is clockwise from the midpoint of the corresponding side. We can also locate a triangle counterclockwise from the midpoint of the corresponding side. I illustrate the sets of cuts on the left of Figure 19.3, with the set from Figure 19.1 in short dashes and the other set of cuts in long dashes. By merging together each pair of right triangles along their side of equal length, we get a set of pieces that we can hinge. This approach works only for circumscribed polygons.

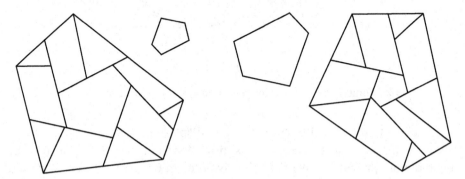

19.4: Hart's hingeable dissection of two circumscribed pentagons to two

The dissection appears in Figure 19.4, applied to the irregular pentagon of Figure 19.1. If the dissection realizes $x^2 + y^2 = z^2 + w^2$ for the circumscribed figure, then the triangles in Figure 19.4 have side lengths in the ratios $x : w : y + z$, where y and z are the side lengths of the two small circumscribed figures. The cyclic hinging, similar to that in Figure 19.2, appears on the right in Figure 19.3. This dissection is also hinge-snug and grain-preserving. How wonderful that even irregular polygons can swing so beautifully!

Turnabout 5: Piano-Hinged Polyhedra

In Turnabout 2, we observed that a triangle built with rods is rigid but a square is not, unless we brace it with additional rods. Suppose that we consider a similar construction for polyhedra. We assume that each face of a polyhedron is rigid but that we use flexible glue on the edges. Alternatively, we could use piano hinges rather than the glue. We ask whether every piano-hinged polyhedron is rigid.

Augustin Louis Cauchy (1813) proved that convex polyhedra are rigid, although his proof needed a patch, supplied by Ernst Steinitz (see Steinitz and Rademacher 1934). The French engineer Raoul Bricard (1897) showed that a structure constructed out of edges, rather than faces, need not be rigid. He designed a non-rigid structure with twelve edges meeting at six vertices. Unfortunately, if we add the eight triangular faces that span between the edges, then several of the faces penetrate each other so that we do not have a simple polyhedron.

Further work on the rigidity question suggested that it might be resolved in either direction. On the one hand, Walter Wunderlich (1965) of Vienna produced a net that specified mountain and valley folds and could be folded into two distinct polyhedra. Michael Goldberg (1978) of Washington, D.C., described two more such nets. On the other hand, Herman Gluck (1975), at the University of Pennsylvania, showed that the fraction of nonconvex polyhedra that are nonrigid is vanishingly small.

Robert Connelly (1978b), at Cornell University, figured out how to modify Bricard's construction so that no two faces penetrate each other while still retaining the polyhedron's flexibility. This necessarily increased the number of vertices and faces. Klaus Steffen (1978), at the University at Düsseldorf, found a simpler polyhedron with only fourteen faces and nine vertices. Indeed, Russian mathematician I. G. Maksimov (1995) proved that any polyhedron with triangular faces and fewer than nine vertices must be rigid.

Connelly (1979) stated two conjectures regarding flexible polyhedra. The first is that if a polyhedron flexes then its volume remains constant. The second conjecture is that, if a given polyhedron flexes from some particular solid to some other particular solid, then there is a dissection in a finite number of pieces from the first solid to the second. Clearly the second conjecture presupposes the truth of the first. Connelly called the first conjecture the "bellows conjecture" because it is equivalent to claiming that a flexible polyhedron will not perform like a bellows.

Connelly, Sabitov, and Walz (1997) proved the bellows conjecture. Their approach was to prove a vast generalization of Heron's formula for the area of a triangle in terms of the lengths of its sides. The second conjecture, about the existence of a finite dissection from one flexing of the polyhedron to another, remains open. Connelly (1978a) also made several other conjectures.

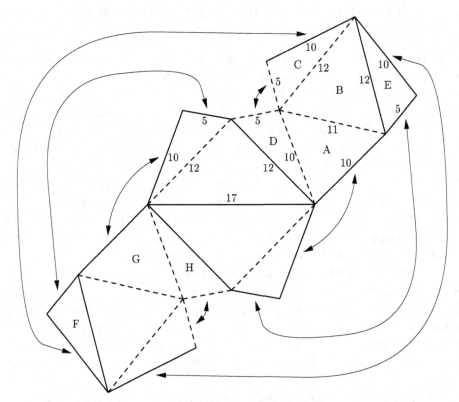

T20: Net for Klaus Steffen's flexible polyhedron

Figure T20 displays the net for Klaus Steffen's polyhedron, with solid edges for mountain folds and dashed edges for valley folds. Arcs match the edges for folding the net into the polyhedron. The net has 2-fold rotational symmetry about the midpoint of the edge of length 17. Lengths label the edges in the top half of the net, with lengths in the bottom half being consistent with the 2-fold symmetry. The letters that label some of the faces help to interpret the next figure. Of course, at least four faces meet at every vertex. Explain why!

Steffen's polyhedron appears in Figure T21 in a perspective view, with the faces labeled in a manner consistent with their labels in the previous figure. We see one complete dimple (faces A, B, C, and D) and part of an identical dimple (faces G and H, and two faces hidden). The edge of length 17 is hidden from view. One configuration of Steffen's polyhedron has 2-fold rotational symmetry. The polyhedron flexes by moving the vertex as shown. Of course, other vertices also move, but the movement is most pronounced for this vertex – the only vertex that maps to itself under the 180° rotation applied to the configuration with 2-fold rotational symmetry.

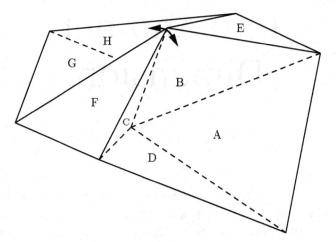

T21: Perspective view of Steffen's polyhedron

CHAPTER 20

Into Another Dimension

We now shift our attention from two-dimensional to three-dimensional figures. Unfortunately, moving into three dimensions seems to add another dimension of difficulty. We know that, even for unhinged dissections, there are pairs of polyhedra such that we cannot dissect one into another in a finite number of pieces. David Hilbert (1900) conjectured this as the third in his famous list of 23 problems, and his student Max Dehn (1900) proved it. Although our techniques for hinging two-dimensional dissections have their analogues in three dimensions, difficulties in visualizing how three-dimensional objects intersect hamper the application of these techniques.

Luckily, one problem – that of dissecting a $2 \times 1 \times 1$ rectangular block to a cube – has attracted considerable attention. American mathematician William Fitch

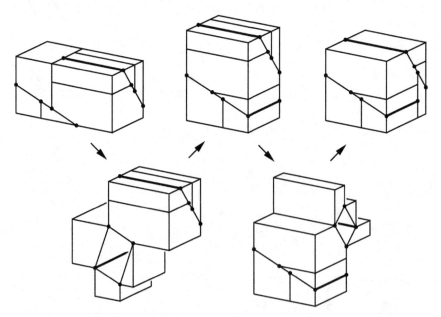

20.1: Hinged dissection of a $(2 \times 1 \times 1)$-rectangular block to a cube

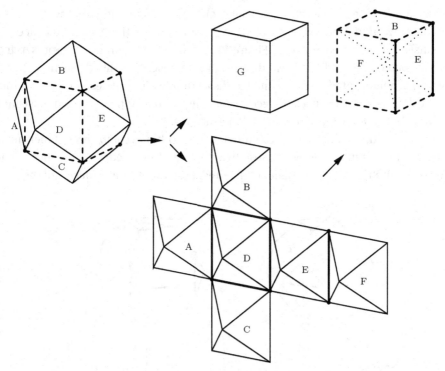

20.2: Hinged dissection of two cubes to a rhombic dodecahedron

Cheney (1933) gave an 8-piece unhingeable dissection of the block to a cube. In (1935), Albert Wheeler, a math teacher in Massachusetts, discovered a 7-piece un-hingeable dissection. Michael Goldberg (1966) posed the problem of finding a hinge-able dissection for this problem and claimed a 10-piece hinged dissection. Anton Hanegraaf constructed a model of a 7-piece cyclicly hinged dissection, a photograph of which appeared in (Tjebbes 1969). Hanegraaf (1989) gave a more detailed discussion, along with a 6-piece unhingeable dissection.

Hanegraaf's dissection unfolds before us in Figure 20.1. The dots and thick lines indicate the portion of the hinges that we can see. The dissection uses two applications of Hanegraaf's trapezoid swing, or T(a)-swing. For that reason, the dissection is grain-preserving. Extending the notion of hinge-snug to piano hinges in three dimensions, we note that the dissection is also hinge-snug. In two dimensions, the hinge-snug property allows us to convert swing hinges to twist hinges. The third dimension makes effective conversion a more difficult problem, to be taken up elsewhere.

Since rhombic dodecahedra honeycomb three-dimensional space and have the same repetition pattern as a pair of cubes, there are simple dissections of a rhombic dodecahedron to two cubes. H. Martyn Cundy and A. P. Rollett (1952) described

199

a very simple hinged dissection. It leaves one cube uncut and cuts the other into six equal pyramidal pieces, each of which covers one of the first cube's faces. It thus uses 7 pieces, as shown in Figure 20.2. This dissection is grain-preserving. Arthur Loeb (1976) and Henk Mulder (1977) also described this dissection. In addition, Loeb illustrated a second hinged dissection, which cuts each cube into four identical parts and thus uses 8 pieces. It also leads to a 10-piece hinged dissection of two cubes to an octahedron and two tetrahedra.

Similarly, the ability of truncated octahedra to honeycomb three-dimensional space leads to simple dissections of two truncated octahedra to a cube. David Paterson (1988) and Anton Hanegraaf noted that there is a 6-piece unhingeable

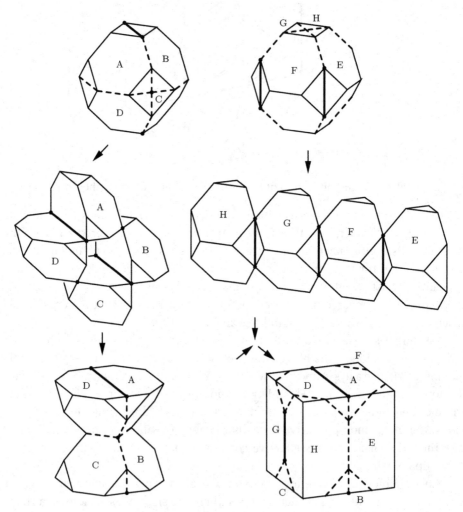

20.3: Hinged dissection of two truncated octahedra to a cube

dissection. Jan Slothouber (1973) gave a 9-piece hinged dissection. Slothouber dissected one of the truncated octahedra into eight identical pieces and cyclicly hinged them so that they could be folded around the other, uncut, truncated octahedron. In my 8-piece hinged dissection (Figure 20.3), I cut each truncated octahedron into four identical pieces. I indicate the hinges with either thick edges, when they are visible along their length, or with dots, when they are visible only at an end. Dashed edges indicate the other boundaries between pieces. An interesting aspect is that I hinge one set of four pieces differently from the other set. Pieces A, B, C, and D are cyclicly hinged, whereas pieces E, F, G, and H are only linearly hinged. This dissection is hinge-snug and grain-preserving.

We have now seen a hinged dissection of two cubes to a rhombic dodecahedron and also two truncated octahedra to a cube. To complete the picture, we should expect a hinged dissection of two rhombic dodecahedra to a truncated octahedron. Indeed, Anton Hanegraaf found a symmetrical 16-piece dissection, once again based on a superposition of honeycombs.

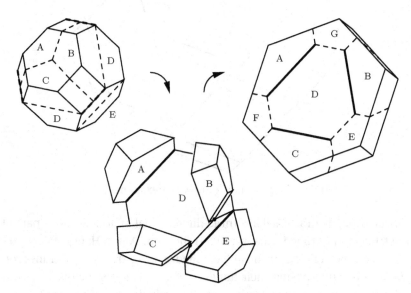

20.4: Truncated octahedron to a hexagonal prism

Anton Hanegraaf also found the nifty symmetrical hingeable dissection of a truncated octahedron to a hexagonal prism in Figure 20.4. He cut the truncated octahedron into pieces on three levels, each of whose thickness is equal to the height of the prism. The middle level (piece D) becomes the center of the prism. The top and bottom levels are each cut into three pieces. Swinging pieces A, B, and C halfway down from the top and piece E halfway up from the bottom gives the intermediate form in the middle of the figure. Completing the turns – and swinging

201

pieces F and G from the bottom – gives the prism on the right. This dissection is hinge-snug and grain-preserving.

Anton derived his dissection by superposing three-dimensional honeycombs. He used the standard honeycomb of the truncated octahedron along with the honeycomb of the hexagonal prism, as started in Figure 20.5. Each layer of hexagonal prisms is offset from the ones above and below it, so that each prism shares precisely three vertices with the prisms beneath it. Also, if a vertex is shared then the corresponding vertex above it or below it in the prism is not shared. Lines of 2-fold rotational symmetry go through the shared vertices.

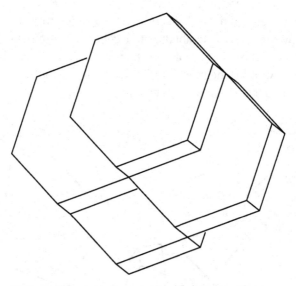

20.5: Building a honeycomb with hexagonal prisms

Truncated octahedra tessellate three-dimensional space. So do a pair of figures: the truncated tetrahedron and the tetrahedron. Anton Hanegraaf discovered that you can dissect a pair of truncated octahedra into a truncated tetrahedron and a tetrahedron by superposing their honeycombs. This gives the dissection in Figure 20.6, with the truncated tetrahedron and tetrahedron on the lower left and the truncated octahedra on the top right and top left. As you can see, the truncated tetrahedron consists of one truncated octahedron (piece A) with a 10-piece hinged assemblage (pieces B through K) wrapped around it. The other truncated octahedron forms when we wrap the hinged assemblage around the tetrahedron (piece L). The vertices of piece L will be in the centers of four of the hexagonal faces of the truncated octahedron. The dissection is hinge-snug.

I give the same hinging as Anton, except that he also hinged pieces G and I together to form a ring of pieces within the assemblage. This makes for a satisfying

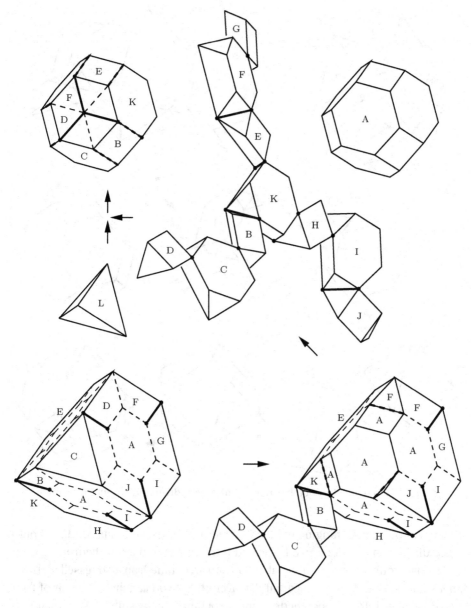

20.6: Two truncated octahedra to a truncated tetrahedron and a tetrahedron

motion when converting his model from one figure to the other. However, I have not been able to verify that the cycle of hinges will work properly if the hinges are precise. My drawing of the hinged assemblage is precise, showing each piece rotated through half of its 180° swing. Readers can see that none of the square and hexagonal faces are in exactly the same plane. Perhaps readers will be able to determine

203

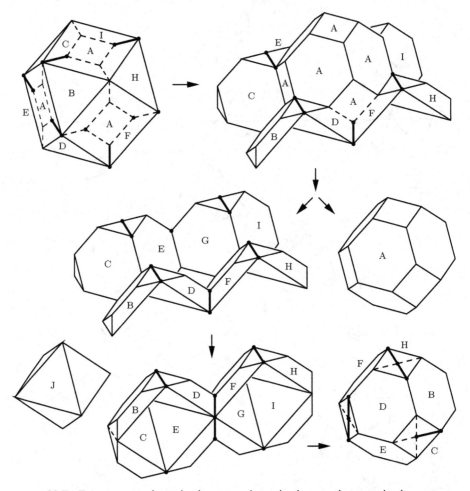

20.7: Two truncated octahedra to a cuboctahedron and an octahedron

whether Anton's cyclic hinging works. Related to this dissection is Anton's 8-piece hinged dissection of a cube to a truncated tetrahedron and a tetrahedron.

The pair consisting of a cuboctahedron and an octahedron also tessellate three-dimensional space. Anton Hanegraaf discovered that you can dissect a pair of truncated octahedra into a cuboctahedron and an octahedron by superposing their honeycombs. This gives the dissection in Figure 20.7, with the cuboctahedron on the upper left, the octahedron on the lower left, and the truncated octahedra on the lower right. The cuboctahedron consists of one truncated octahedron (piece A) with a lovely 8-piece cyclicly hinged assemblage (pieces B–H) wrapped around it. The 8 pieces are identical and symmetric. The other truncated octahedron forms when we wrap the hinged assemblage around the octahedron (piece J). The vertices of piece J are at the centers of of the six square faces of the truncated octahedron. The dissection is hinge-snug and grain-preserving.

In a manner vaguely reminiscent of Figure 20.2, Anton Hanegraaf dissected a truncated octahedron into a stellated rhombic dodecahedron using six hinged pieces. One can imagine slicing six caps off of a truncated octahedron, as suggested by the upper left in Figure 20.8. However, instead of leaving a rectangular solid in the middle, Anton divided this solid to obtain the points of the stellated figure.

David Paterson (1988) gave a 17-piece unhingeable dissection of a truncated octahedron to a cube, and Anton Hanegraaf found a 13-piece unhingeable dissection. Gavin Theobald found a nifty improvement to Hanegraaf's dissection, reducing it to

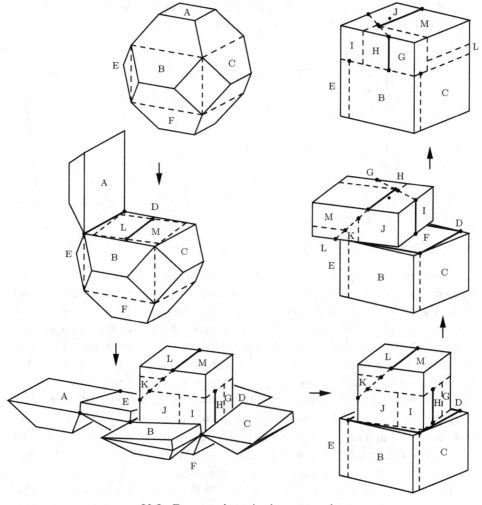

20.8: Truncated octahedron to a cube

12 pieces. Adding my own twist, I found an interesting way to combine Theobald's dissection and Hanegraaf's 7-piece hinged dissection of a rectangular block to another rectangular block, similar to that in Figure 20.1. My 13-piece hingeable dissection first turns, then twists in Figure 20.8.

Following Theobald's dissection, we first cut the cap (piece A) off the truncated octahedron and then cut the side pieces (B, C, D, and E), but we include triangular prisms in those pieces so that pieces A, B, C, D, and E, along with the bottom piece F, form a rectangular block whose length and width equal the side length of the desired cube. We cut the remaining rectangular block from the center of the truncated octahedron in a fashion similar to what we did in Figure 20.1 to give another rectangular block whose length and width each equal the side length of the desired cube. However, this (G–M)-block sits on top of the (A–F)-block in an odd orientation, as shown from above on the left of Figure 20.9.

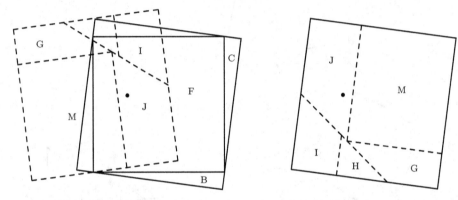

20.9: Top view of the last step of a truncated octahedron to a cube

However, we can get the desired cube by rotating the top block about a vertical axis with a center of rotation in piece J. The center of rotation is shown as a small dot on the top of piece J in the last two parts of Figure 20.8 and the corresponding top views in Figure 20.9. We use a *socket hinge* here, which is the three-dimensional analogue of a twist hinge. We can identify the center of rotation in piece J because of the following property. Choose any point in piece J and locate its positions before and after the rotation. The center of rotation must be equidistant from these two positions, and thus the center of rotation must lie on the perpendicular bisector of the line segment with these two positions as endpoints. If we choose a second point in piece J and locate its positions before and after the rotation, then we can find a second perpendicular bisector. Intersecting the two bisectors gives the center of rotation. All of the piano hinges are hinge-snug.

When Hanegraaf saw my 13-piece hingeable dissection of a truncated octahedron to a cube, he responded by finding a 14-piece dissection (Figure 20.10) that

206

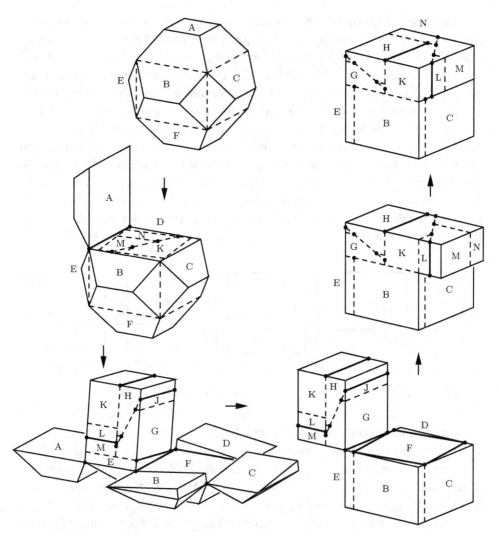

20.10: Hanegraaf's truncated octahedron to a cube

uses only piano hinges. He also used Theobald's approach so that pieces A, B, C, D, and E, along with the bottom piece F, form a rectangular block whose length and width equal the side length of the desired cube. However, he differed in how he cut the remaining rectangular block from the center of the truncated octahedron.

If you fold over this remaining block using a piano hinge, then you cannot use the 7-piece conversion technique borrowed from Figure 20.1. This is because no corner remains a corner after you perform two T(a)-swings. Instead, Hanegraaf replaced one of the T(a)-swings with a Q-swing, which converts the block consisting of pieces G, H, I, and J to a different block. (Pieces I and J are relatively small and

207

are not labeled everywhere.) The Q-swing preserves two edges as external when it converts one block to another, and Hanegraaf used both of these edges to attach piano hinges. The final T(a)-swing converts a block consisting of pieces K, L, M, and N to an L-shaped piece that fills out the cube. Using similar ideas, Anton Hanegraaf also found a 15-piece hingeable dissection of a rhombic dodecahedron to a cube.

Robert Reid found an 8-piece unhingeable dissection of cubes for $1^3 + 1^3 + 5^3 + 6^3 = 7^3$. Robert's superb dissection is clearly minimal among rational dissections of those cubes. However, an 8-piece hingeable dissection seems to be too much to hope for. I have found a 10-piece hingeable dissection, which I give in Figures 20.11 and 20.12. As in Robert's dissection, I leave the two 1-cubes and the 5-cube uncut. I cut the 6-cube into three larger rectangular blocks A, C, and E, which fill in against three of the faces of the 5-cube (piece H), and four smaller rectangular blocks B, D, F, and G.

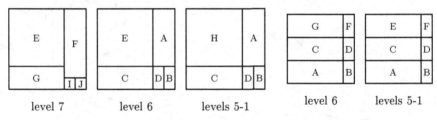

20.11: Hingeable dissection of cubes for $1^3 + 1^3 + 5^3 + 6^3 = 7^3$

I then apply a technique from Figure 7.1 that I used for cutting a 4-square into a hinged assemblage of three rectangles (a 2×4, a 2×3, and a 1×2) that form a 5-square minus a 3-square. The first application of the technique involves pieces A, B, C, and D, where A plays the same role as 2×3, B plays the same role as 1×2, and C and D together play the role of 2×4. The second application involves pieces C, D, E, F, and G, where E and F together play the role of 2×3, G plays the role of 1×2, and C and D together play the role of 2×4. The reason why C and D are not just one piece is that, if they were, then A and B would open downward in levels 1–6 of the 7-cube in Figure 20.11. Thus, I cut piece D from piece C and connect them with a socket hinge whose axis goes through the centers of each piece. Then I rotate piece D by 180° relative to piece C. The reason to split pieces E and F is that otherwise pieces E and A would both compete for a $(5 \times 1 \times 1)$ portion of the 7-cube.

To allow piece D to rotate, I use a ball-and-socket joint to connect pieces G and F. The joint connects G and F at the point on the boundary of pieces E, F, and G on the exterior of the 7-cube. In the 6-cube, the joint connects G and F at the point on the boundary of pieces C, D, E, F, and G. To move from the 6-cube to the 7-cube, I rotate pieces G and F by 180° so that G is flush against piece C. Then I rotate piece

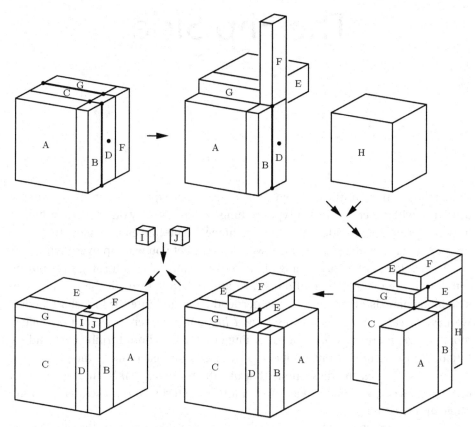

20.12: Hinged transformation of cubes for $1^3 + 1^3 + 5^3 + 6^3 = 7^3$

F by 90° relative to piece G so that piece F does not obstruct the rotation of piece D. Then I rotate pieces A and B so that A is flush with both B and D. Next, I rotate D (bringing along A and B) 180° relative to C. Then I rotate F by 180° to bring it flush on top of A. Finally, I slide in pieces E, I, and J to complete the 7-cube.

Once we allow ball-and-socket joints, are we perilously close to being unhinged? There is a 34-piece hingeable dissection of a dodecahedron, icosahedron, and icosidodecahedron to a small stellated dodecahedron and a large stellated dodecahedron. Just slice the icosidodecahedron by six planes, each of which goes through the center of the icosidodecahedron and contains ten of its edges that form a regular decagon in that plane. Then attach the resulting 20 triangular pyramids to the faces of the icosahedron and the resulting twelve pentagonal pyramids to the faces of the dodecahedron. Since illustrating this dissection is in another dimension of difficulty, let's quickly move on to the next chapter. There we will experience a more limited interaction with the third dimension.

CHAPTER 21

The Flip Side

Hinging dissections of three-dimensional objects has expanded our minds, forcing us to deal with several different types of hinges. Now let's try other types of hinges in two dimensions, in addition to swing hinges. The hinges will allow us to flip pieces over and in many cases use fewer pieces. I bet you will flip over them.

The analogue of a socket hinge in two dimensions allows turning over but no rotation in the plane. The motion is a rotation of 180° through the third dimension along an axis of rotation that is perpendicular and interior to a shared edge. We have already identified this as a *twist hinge* in Chapter 1. We can also allow more general motion. A *flip hinge* is a hinge that allows both turning over and rotation. Consequently, it must appear at a vertex of at least one of the pieces that it connects. We call the dissections that use them *flip-hingeable.* In drawing a set of hinged pieces, we shall use a dot with a white interior to represent either a flip hinge or a twist hinge.

Sam Loyd (*Tit-Bits*, 1897c) posed the problem of dissecting a clipped rectangle to a square in three pieces. We see Loyd's "clipped rectangle" in Figure 21.1. The original rectangle is twice as high as it is wide, and the triangle is clipped off it with a cut from the midpoint of the right side at an angle of 15° from the vertical. Dotted lines indicate the removed triangle.

In Figure 21.2 I mark the piece that we turn over with an "∗" in the clipped rectangle and with a "⋆" in the square. One cut starts at the midpoint of the slanted

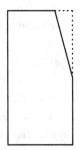

21.1: Clipped rectangle

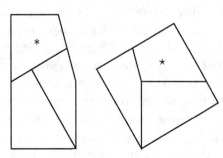

21.2: Loyd's clipped rectangle to square

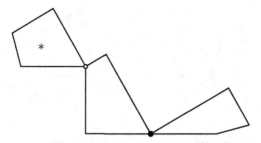

21.3: Swing and flip hinges for a clipped rectangle to a square

side and is equal in length to the width of the clipped rectangle. I swing- and flip-hinge the pieces in Figure 21.3. Neither Loyd nor "Sphinx" (Henry Dudeney, who wrote up the solutions) identified the dissection as swing-and-flip-hingeable.

Dudeney did state that the angle need not be restricted to 15°. Harry Lindgren (1970) showed that the solution technique will work for angles up to 22.5°, but only if the dimensions of the original rectangle are also changed. If the angle is α then the height of the rectangle is $(1 + 2\sin(2\alpha))$ times its width, and the clip begins at a point above the base at a height equal to the rectangle's width. The dissection is swing-and-flip-hingeable over this entire range of α. The pieces connected by the swing hinge are hinge-snug.

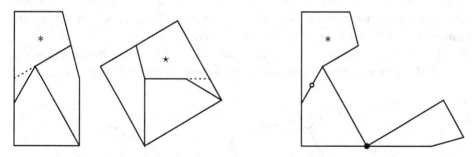

21.4: Swing-and-twist-hinged variation of Loyd's clipped rectangle to a square

A flip hinge is a rather general mechanism and is sometimes more than we need. We can modify the dissection of a clipped rectangle so that a twist hinge replaces the flip hinge. We just slice an isosceles triangle off one piece and glue it onto the other of the two pieces connected by the twist hinge. We then position the twist hinge at the midpoint of the base of the isosceles triangle. We can admire the modified dissection in Figure 21.4.

Whenever a flip hinge connects just two pieces, we can replace the flip hinge by a swing hinge and a twist hinge at the expense of using one more piece. Thus

211

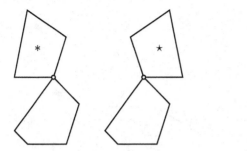
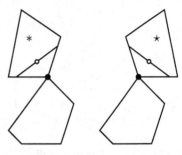

21.5: Converting a flip hinge ...   **21.6:** ... to swing and twist hinges

we can replace the flip-hinged assemblage of Figure 21.5 by the assemblage in Figure 21.6. The trick is to slice an isosceles triangle off one of the two pieces so that the two equal sides of the isosceles triangle are incident on what was the flip hinge.

Some improvements in the number of pieces are relatively easy to achieve when we allow the use of flip and twist hinges in addition to swing hinges. In Chapter 5, we have seen a 10-piece hingeable dissection of two unequal hexagons to one in Figure 5.7. Do you see how to merge the three pieces of the "gadget" together into one piece and then to connect that merged piece to the piece on its left with a twist hinge, giving an 8-piece dissection?

What other dissections can we improve, if we use flip and twist hinges as well as swing hinges? We'll take a quick sampling, as long as *you* don't flip out! Whenever possible, we will choose twist hinges instead of flip hinges, anticipating the wholesale use of twist hinges in the next chapter.

Alfred Varsady discovered an 18-piece dissection of five hexagons to one, which I modified to give a 16-piece dissection. Then Anton Hanegraaf found a nice 15-piece

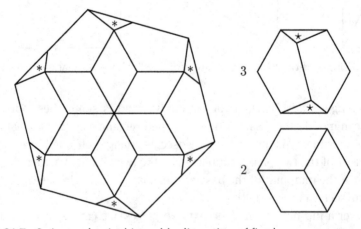

21.7: Swing-and-twist-hingeable dissection of five hexagons to one

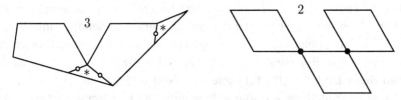

21.8: Swing and twist hinges for five hexagons to one

dissection. But to find a hingeable dissection, we return to Alfred's original dissection, which Anton identified as swing-and-flip-hingeable. In Figure 21.7, I modify Alfred's original dissection so that we can use twist hinges to turn over 6 pieces. To achieve this, we cut two isosceles triangles from neighboring pieces and merge them with each triangle that Alfred turned over. The hinged pieces use nine twist hinges, as we see in Figure 21.8. Each of the twist hinges is midway along one of the sides of the triangles resulting from the merges. We fill in the hexagram interior of the large hexagon by using the swing-hingeable (and hinge-snug) dissection from Figure 3.25.

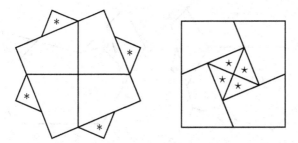

21.9: Swing-and-twist-hingeable dissection of an {8/2} to a square

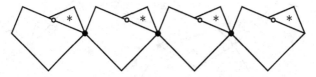

21.10: Swing and twist hinges for an {8/2} to a square

The 8-piece dissection of an {8/2} to a square that appeared in the circa 1300 Persian manuscript, *Interlocks of Similar or Complementary Figures* – and that Harry Lindgren (1964b) rediscovered – is closely related to the dissection of two unequal squares to one (Figure 4.1) that Henry Perigal (1873) published. The dissection has 4-fold rotational symmetry but uses one more piece than my (1972d) dissection. No one previously realized that this dissection (Figure 21.9) is swing-and-twist-hingeable (Figure 21.10). It is an alternative to the 11-piece swing-hingeable dissection of Figure 15.15. Each of the twist hinges is midway along one of the sides of the four larger pieces. The dissection is grain-preserving, and the pieces connected by swing hinges are hinge-snug. The same trick gives an 8-piece swing-and-twist-hingeable dissection of a Greek Cross to an {8/2}.

When Harry Lindgren (1964b) came to dissect a {12/2} to a hexagon, he discovered that he could base a 10-piece dissection on his dissection of two unequal hexagons to one, which is implied by the superposition in Figure 5.6. It has two more pieces than my (1972d) dissection. Again, Harry apparently did not realize that his dissection (Figure 21.11) is swing-and-twist-hingeable, as I illustrate in Figure 21.12. Using the twist hinges allows us to get away with two fewer pieces than the purely swing-hingeable dissection in Figure 10.21. Each of the twist hinges is midway between a convex and a concave vertex of the {12/2}. The pieces connected by swing hinges are hinge-snug.

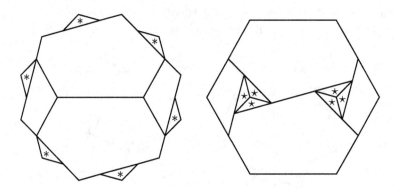

21.11: Swing-and-twist-hingeable dissection of an {12/2} to a hexagon

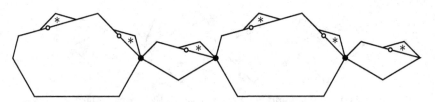

21.12: Swing and twist hinges for a {12/2} to a hexagon

Lindgren's dissection implicitly converts the {12/2} to a 7-piece tessellation element. We can substitute this element in place of the 9-piece element in Figure 10.20 for other dissections using that element. We can thus discover that a symmetric variation of Stuart Elliott's 13-piece dissection of a {12/2} to three hexagons (see p. 179) is swing-and-twist-hingeable. Those pieces that connect via swing hinges are hinge-snug. Also, we can substitute an 8-piece strip element for a {12/2} in place

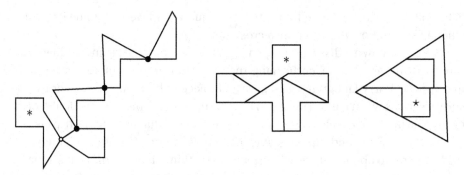

21.13: Swing-and-flip-hingeable dissection of a Latin Cross to a {3}

of the 10-piece element that we use in Figure 11.52. This gives a 12-piece swing-and-twist-hingeable dissection to a square whose swing-hinged pieces are hinge-snug.

We come now to dissecting a Latin Cross to a triangle. Lindgren (1964b) gave a 5-piece unhingeable dissection, and we have already seen a 7-piece swing-hingeable dissection in Figure 11.50. The 2-piece penalty is a bit frustrating, since all pieces in Lindgren's dissection except for one small square hinge together. To find a swing-and-flip-hingeable dissection, Anton Hanegraaf adapted Lindgren's dissection, cutting an isosceles triangle out of one piece and gluing the small square to it. By flipping the isosceles triangle around on its axis of reflection symmetry, Hanegraaf then brought the square into position in the other figure. This gives the 5-piece dissection pictured in Figure 21.13. We can modify this dissection in the same manner as in Figure 21.4 to replace the flip hinge by a twist hinge.

Anton's adaptive approach can be classified as a bumpy crossposition. The dissection uses the crossposition in Figure 21.15, which takes a Latin Cross element as produced in Figure 21.14. The T-strip for the Latin Cross is a bumpy strip, with squares poking up and down from it, and with corresponding indentations. There

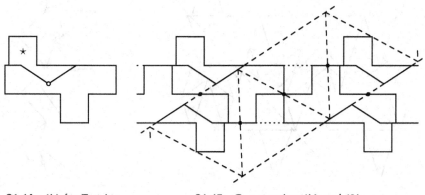

21.14: {L} for T-strip   **21.15:** Crossposing {L} and {3}

215

is flexibility in exactly where to position the square, and we determine its final position based on how the other strip crosses it.

We come now to dissecting a pentagram to a square. Martin Gardner (1958) proposed the problem of performing the dissection into no more than 9 pieces; in response, Harry Lindgren (1958) gave an 8-piece unhingeable dissection. What a surprise it was when Philip Tilson (1978) came up with a 7-piece dissection! More recently, Gavin Theobald also found a 7-piece dissection that uses his T-strip element.

To find a swing-and-flip-hingeable dissection, Anton Hanegraaf adapted Theobald's T-strip to obtain a 9-piece hingeable dissection. I have modified his approach by using a bumpy crossposition to yield the 9-piece swing-and-twist-hinged dissection of Figures 21.17 and 21.19. I cut a pentagram as in Figure 21.16 to form a T-strip element, which I use in a crossposition that challenges us in Figure 21.18. The T-strip for the pentagram is a bumpy strip, with isosceles triangles poking up and down from it and with corresponding indentations. I position these triangles using rotation about twist hinges.

21.16: {5/2} for T-strip

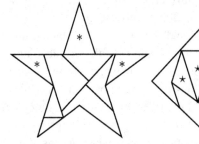

21.17: Hingeable {5/2} and square

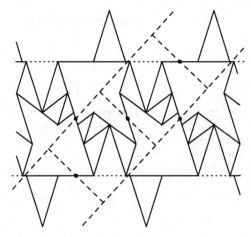

21.18: Crossposing {5/2} and square

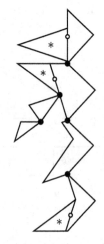

21.19: Hinged pieces

Lindgren (1964b) also gave a 10-piece unhingeable dissection of a pentagram to a hexagon. Again, Gavin Theobald showed no mercy, finding a 9-piece dissection based on his T-strip element. And once again flipping out, Anton Hanegraaf found an 11-piece swing-and-flip-hingeable dissection closely related to his dissection of a pentagram to a square. The same modification that I brought to bear on that dissection works here, giving an 11-piece swing-and-twist-hinged dissection (Figures 21.21 and 21.23). Using the T-strip element from Figure 21.20, we get a bumpy crossposition to twist our grey matter in Figure 21.22. One of the isosceles triangles is almost flipped on its axis of reflection symmetry. If we merge the two pieces linked by the twist hinge, we get an approximate 10-piece dissection.

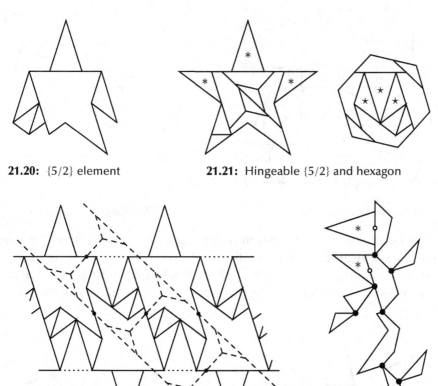

21.20: {5/2} element **21.21:** Hingeable {5/2} and hexagon

21.22: Crossposing {5/2} and hexagon **21.23:** Hinged pieces

Let's proceed to a dissection of a Maltese Cross to a Greek Cross. Bernard Lemaire's dissection, given by Berloquin (*Le Monde,* 1975a), turns out to be swing-and-flip-hingeable. We can modify it by changing the flip hinges to twist hinges as in Figure 21.25. First we twist four pieces from the Maltese Cross to produce

217

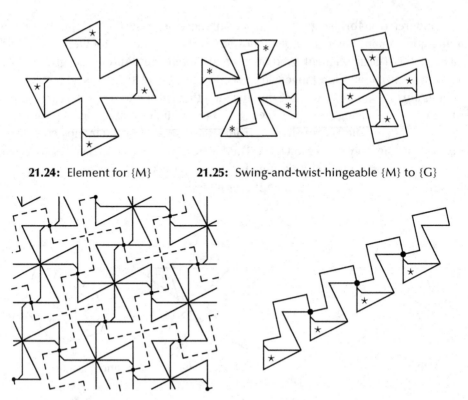

21.24: Element for {M}   **21.25:** Swing-and-twist-hingeable {M} to {G}

21.26: Tessellations for {M}, {G}   **21.27:** Hinges for {M}, {G}

the tessellation element that twists in Figure 21.24. Then we superpose the tessellation of these elements with a tessellation of Greek Crosses, as in Figure 21.26. The dissection in Figures 21.25 and 21.27 will work with any Maltese-like cross in which the distance between outer vertices of neighboring arms equals the distance

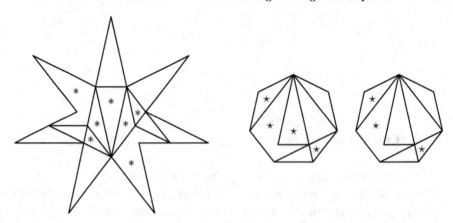

21.28: Swing-and-twist-hingeable dissection of two heptagons to a {7/3}

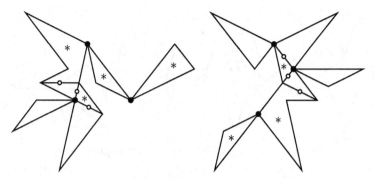

21.29: Swing and twist hinges for two heptagons to a {7/3}

between opposite inner vertices. I have not shown the twist hinges at the midpoints of the four short diagonal edges. The swing-hinged pieces in this dissection are hinge-snug.

An especially startling dissection is of two heptagons to a {7/3}. In (1972a) I gave a 9-piece unhingeable dissection, and I base my 14-piece swing-and-twist-hingeable dissection in Figures 21.28 and 21.29 on the same approach. However, to correctly orient some of the pieces, I cut a triangle into smaller triangles and use twist hinges to flip the pieces around appropriately. Although I cut both heptagons identically, I hinge them slightly differently.

In a fashion similar to swing hinges, we can allow more than two pieces to be flip-hinged together at a point and can even allow a combination of a swing and a flip hinge. Be ready for such a hybrid to flip your wig when we dissect four pentagrams to one.

Harry Lindgren (1964b) attributed to Ernest Freese a 14-piece unhingeable dissection of four pentagrams to one, but this was bettered by Stuart Elliott (1985) with

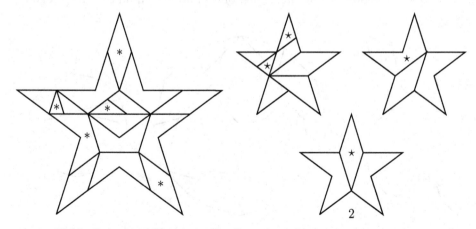

21.30: Swing-and-flip-hingeable dissection of four pentagrams to one

219

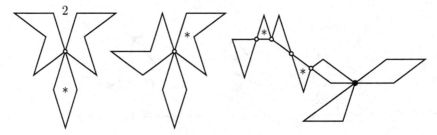

21.31: Swing and flip hinges for four pentagrams to one

a 12-piece unhingeable dissection. Using swing hinges only, the best hingeable dissection I have been able to manage has more than three times as many pieces. With flip as well as swing hinges, I found a 17-piece dissection by adding three more cuts Freese's basic approach. Anton Hanegraaf then saw how to do it by adding only two more cuts; see Figures 21.30 and 21.31.

Freese's 8-piece dissection of two octagons to one turns out to be swing-and-twist-hingeable.

Puzzle 21.1 Describe an 8-piece swing-and-twist-hinged dissection of two octagons to one.

We have already seen Anton Hanegraaf's 20-piece swing-hingeable dissection of three dodecagons to one in Figure 17.13. Starting from a symmetrical 15-piece unhingeable dissection of mine, Robert Reid found a symmetrical 18-piece swing-and-twist-hinged dissection. Since his variation was not hinge-snug, I looked for an alternative one that is hinge-snug. The result, in Figure 21.32, is very close to my original unhingeable dissection in that I make just one additional cut in each dodecagon.

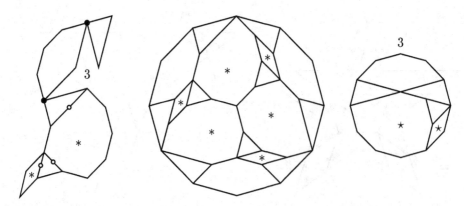

21.32: Swing-and-twist-hinged dissection of three dodecagons to one

We discover that twist hinges also help in dissecting three octagons to one. We save a piece over the 14-piece dissection of three octagons to one in Figure 17.19. The swing-and-twist-hinged dissection in Figures 21.33 and 21.34 has an attractive 2-fold rotational symmetry. One pair of twist hinges allows us to rotate a piece a quarter-turn, and another allows us to flip a pair of rhombuses out of a small octagon. The swing-hinged pieces are hinge-snug.

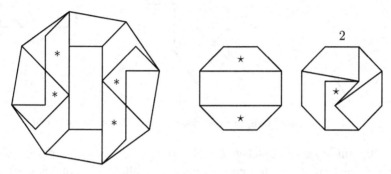

21.33: Swing-and-twist-hingeable dissection of three octagons to one

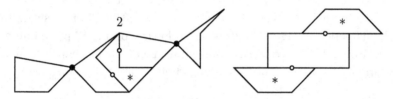

21.34: Swing and twist hinges for three octagons to one

We have already seen a 25-piece hingeable dissection of three {8/2}s to one in Figure 17.21. If we also use twist hinges, we need only 17 pieces for the dissection in Figure 21.35. The idea is to leave one small octagram uncut and to cut the other two symmetrically. In fact, this is the original dissection that I found before discovering the 16-piece one (Frederickson 1974). The only trick necessary is recognizing that we can rotate the small isosceles right triangles 180° to achieve their repositioning. The point of rotation lies halfway between one of the sharp points of the isosceles right triangle and a vertex of the octagram. The hinged pieces show the isosceles right triangles already flipped. The four-sided pieces are held in place by the triangles and cannot be swung on their hinges until we twist at least some of the triangles out of the way. The dissection as shown is not hinge-snug, but it can be made so by shifting an appropriate isosceles trapezoid from each isosceles right triangle to its neighboring quadrilateral. The twist hinge would then be on the top edge of the trapezoid.

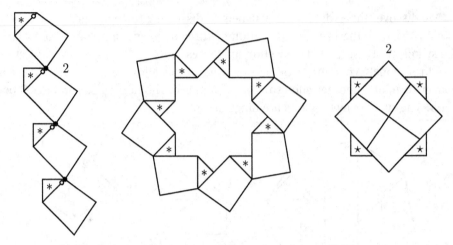

21.35: Swing-and-twist-hinged dissection of three {8/2}s to one

Now it's time to give ourselves a real "lift." In Figure 17.29 I gave a 25-piece hingeable dissection of five decagons to one. If we allow a flip-hinge, then we can do better. Actually, the flip hinge is needed not to flip any pieces over but rather to lift a piece up and out of the way while another piece is rotated around. Anton Hanegraaf has called this type of hinge a *lift hinge.* The 20-piece swing-and-lift-hingeable dissection in Figure 21.36 is based loosely on my 17-piece unhingeable dissection (1974). The three pieces attached to the lift hinge change their relative ordering around the hinge. The swing-hinged pieces are hinge-snug.

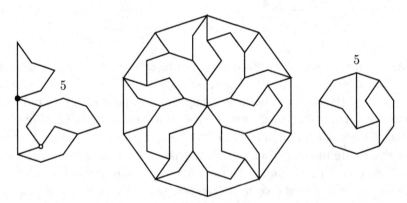

21.36: Swing-and-lift-hinged dissection of five decagons to one

Whenever a lift hinge connects just three pieces, we can replace the three pieces by four pieces and the lift hinge by three twist hinges. Thus we can replace the assemblage in Figure 21.37 by the assemblage in Figure 21.38. The trick is to identify isosceles triangles on each of the three pieces so that they share sides of the same length. We then slice these triangles off their respective pieces and glue them together to get the fourth piece. Suppose that one of the angles incident on the lift hinge is at least 180°, as in the last figure. In this case, cut three identical isosceles triangles out of the one piece and then glue them together with the isosceles triangles from the other two pieces.

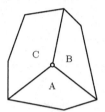 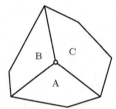  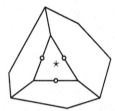

21.37: Converting a 3-piece lift …   **21.38:** … to twist hinges

Alfred Varsady surprised many people by finding $\sqrt{7}$ as one of the dimensions in the heptagon. This discovery led Varsady to a 35-piece unhingeable dissection, which I improved to a 21-piece unhingeable dissection. Anton Hanegraaf took my 21-piece dissection and – using swing and flip hinges – found a 28-piece dissection. He split one piece in each small heptagon into two pieces, one of which is flipped to provide the connection to a remaining piece. I have modified my 21-piece unhingeable dissection to the 28-piece swing-and-twist-hinged dissection taking center stage in Figure 21.39. The trick is to cut an isosceles triangle off one piece in each small heptagon and then use two twist hinges to move an adjacent piece into the proper position. The swing-hinged pieces are hinge-snug.

Since these are heptagons rather than octagons, we can't stop now! In Chapter 17, I gave a 36-piece swing-hingeable dissection of eight heptagons to one. Anton

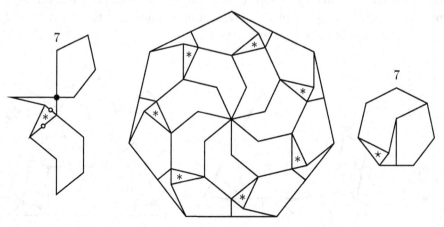

21.39: Swing-and-twist-hinged dissection of seven heptagons to one

223

Hanegraaf identified a 29-piece swing-and-flip-hingeable dissection, and after a while I found a 29-piece swing-and-twist-hingeable dissection (Figure 21.40). Starting with my 22-piece unhingeable dissection, I slice an isosceles triangle off of one of the pieces in each small heptagon. Flipping the isosceles triangle around transports the turned-over piece into the right position. Both swing hinges abut each other in the hinged piece; I have indicated their separation by drawing a white bar between them.

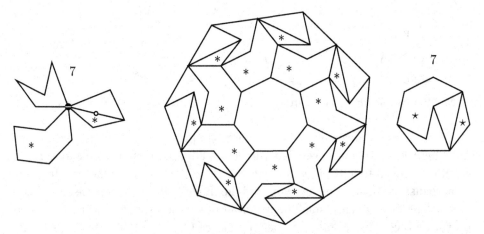

21.40: Swing-and-twist-hinged dissection of eight heptagons to one

The flips that we have seen here are anything but flops. Besides saving pieces with their turnover motion, they anticipate the fully twist-hingeable dissections of the next chapter. However, Anton Hanegraaf found quite a few swing-and-flip-hingeable dissections that do not readily transform into twist-hingeable dissections. Some of the nicest of his dissections in this category are a 10-piece dissection of a heptagon to a triangle, an 11-piece dissection of an enneagon to a triangle, and a 10-piece dissection of a decagon to a triangle.

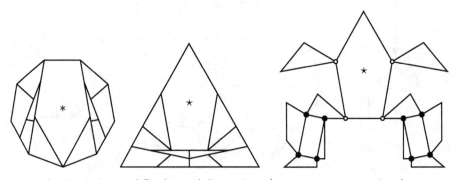

21.41: Swing-and-flip-hinged dissection of an enneagon to a triangle

Let's enjoy Anton's dissection of an enneagon to a triangle, which is shown in Figure 21.41. He sliced two isosceles trapezoids off the enneagon and then applied T(a)-swings to them. They don't look like regular T(a)-swings because the figures formed in the triangle are not trapezoids, yet they are T(a)-swings on somewhat larger trapezoids from which a portion has been removed. Thus Anton was able to fill in below the top part of the triangle. Those top pieces come naturally from the enneagon, and they were cut and flip-hinged so that the pieces from the trapezoids could be hinged to them. With this, are you ready for blast-off?

Curious Case, part 6: Statistically Piquing

Who was C. W. McElroy? Was he so good that he could have discovered the 4-piece dissection that Henry Dudeney might have missed? Let's gather some statistics that will perhaps pique our curiosity. And while we are collecting statistics, let's see when Dudeney granted extensions to his readers.

Charles William McElroy was born on August 25, 1861. He was the son of William McElroy, a mercantile clerk who lived in Chorlton upon Medlock, near Manchester, England. By the time that he was 20, Charles McElroy was working as a bank clerk in Manchester. Later in his life he was a clerk for a metals broker. He died in Manchester in 1928.

The table (Figure C5) shows McElroy's performance on puzzles in the *Weekly Dispatch*. For a given number of solvers, we have the number of times that McElroy was one of the solvers, the number of times that there were that number of solvers, and McElroy's percentage. Dudeney did not always give the list of correct solvers,

Number of solvers	Count for McElroy	Count for everyone	Percentage McElroy
1	3	23	13.0
2	3	18	16.6
3	5	6	83.3
4	8	11	72.7
5	5	7	71.4
6	8	10	80.0
7	4	9	44.4
8	3	9	33.3
9	4	6	66.6
10	3	5	60.0
11	4	4	100.0
12	6	7	85.7
13	4	4	100.0
14	3	4	75.0
15	5	6	83.3
16	4	4	100.0
17	5	5	100.0
18	4	6	66.6
19	5	5	100.0
20–29	21	23	91.3
30–39	11	13	84.6
40–49	3	5	60.0
over 49	3	9	33.3

C5: McElroy's performance quantified by the number of solvers for puzzles during the six-year period 1898–1903

either because there were too many or because he lacked space. Over the six-year period, an average of about eleven puzzles per year had no list of correct solvers. In addition, at least eighteen puzzles over the six-year period were not mathematical and thus were not included in the counts.

Dudeney listed McElroy as a correct solver on over half the math puzzles in which correct solvers were listed. Since hundreds of readers typically submitted solutions, McElroy was clearly a member of what Dudeney termed his "band of experts." No one else matched McElroy as a three-time single-solver. Only one other person had been one of two solvers at least three times: George Wotherspoon, who had earned an M.A. in mathematics from Oxford, achieved that honor four times. Thus, McElroy was arguably the best solver and clearly capable of discovering a solution that Dudeney had missed.

Under what circumstances did Dudeney grant or not grant extensions to the readers of *The Weekly Dispatch*? For "The Scientific Skater" on January 9, 1898, and "Pigs in Pens" on November 21, 1897, he granted extensions to clarify conditions. For "The Fleur-de-Lys Puzzle" (July 4, 1897), "The Gardener and the Cook" (April 15, 1900), "Solitaire Dominoes" (February 3, 1901), "Magic Squares of Two Degrees" (February 22, 1903), and "The Great Cigar Puzzle" (June 21, 1903), Dudeney gave competitors an extra two weeks when no one had sent in the solution.

Aside from the "Triangle and Square," Dudeney did not grant extensions when there was one solver, as demonstrated by "The Eleven Bears" (August 20, 1899), "A Disley Puzzle" (August 3, 1902), and "The Spider and the Fly" (June 28, 1903). It is curious that, in *The Canterbury Puzzles*, Dudeney cited McElroy as the only solver for the triangle-to-square puzzle but did not do the same for "The Eleven Bears."

In "The Spider and the Fly," Dudeney implicitly stated the rule that there would be no extension if anyone solved the puzzle: "It is remarkable that only one correspondent out of a tremendous number who sent in their answers was able to see that there is any shorter route than the the 42ft. path that I have first mentioned.... The correct answer was only received from Mr. J. W. Dippy ... Mr. Dippy thus saves me from having to repeat an unsolved puzzle two weeks in succession." However, for "The Table-Top and Stools" (September 7, 1902), no one found the intended solution yet Dudeney gave no extension.

Thus, Henry Dudeney's extension for the "Triangle and Square" was clearly anomalous. What do you think of our famous puzzlist now?

CHAPTER 22

Twist and Shout!

As I discovered and compiled dissections for the previous chapter, I became increasingly fascinated with twist hinges. A twist hinge is a truly fundamental type of hinge in that it allows only one alternative: a rotation of 180° through the third dimension. Twist hinges are thus remarkably economical in comparison with flip hinges. Motivated by this observation, I substituted twist hinges for flip hinges whenever possible in Chapter 21. I wondered if I could always replace the flip hinges by (perhaps more) twist hinges.

A more intriguing question almost escaped me: Can we dissect any set of figures to another of equal area so that we can hinge them using only twist hinges? At first, this question seemed improbable and ill-considered. How could we possibly get sufficient generality with this rather basic hinge? I did not seriously consider using only twist hinges until I saw two marvelous dissections by Bernard Lemaire.

Bernard has investigated many different types of crosses and dissected most of them to squares. In August 1999, he showed me a dissection (Figure 22.1, but without the hinges) of a square and his $\{\check{M}'_3\}$ cross. This cross is related to $\{\check{M}\}$, which we have seen in Figure 12.13. The outline of $\{\check{M}'_3\}$ looks tricky enough in Figure 22.2. I was surprised to discover that the dissection is twist-hingeable, a fact that Bernard himself had not noticed.

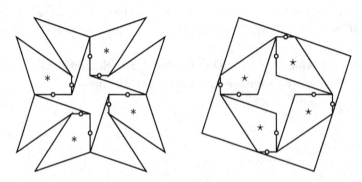

22.1: Twist-hinged dissection of $\{\check{M}'_3\}$ to a square

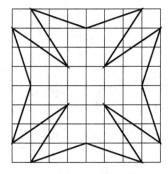

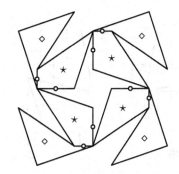

22.2: Outline of {M̌′₃}   **22.3:** Intermediate configuration

We see an intermediate configuration of the pieces, showing the position of the twist hinges, in Figure 22.3. We can produce it from the cross by twisting the four pieces that are marked by the asterisks in Figure 22.1. Since each of these pieces has another piece attached to it, we actually turn over pairs of pieces. Twisting back the other pieces, marked by diamonds, then gives the square.

At the same time, Bernard showed me a dissection (Figure 22.4) of a square and his {M̌′₅} cross. This cross (Figure 22.5) is also related to {M̌}. The surprises continued with this dissection, which can be hinged using only twist hinges – again, Bernard had not noticed this. This second dissection and its hinging are subtly different from the first.

Compare the intermediate configuration for the last dissection with the intermediate arrangement for this one (Figure 22.6). We can produce it from the cross by twisting the four pieces that are marked by the diamonds in Figure 22.6. Twisting pairs of pieces, each marked by an asterisk and a diamond, then gives the square.

The twist-hingeable examples from Bernard are just magical. What a wonderful motion! First twist,... *then shout!* But Bernard had supplied no method for finding

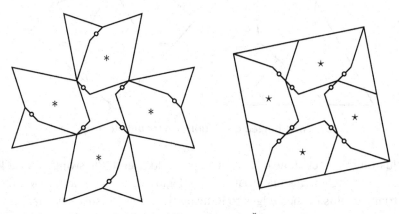

22.4: Twist-hinged dissection of {M̌′₅} to a square

229

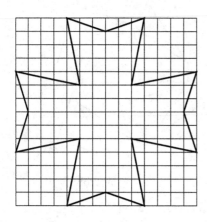

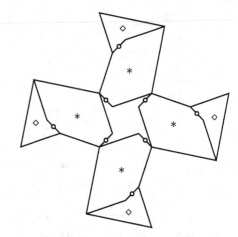

22.5: Outline of $\{\check{M}'_5\}$ **22.6:** Intermediate configuration

them. Fortunately, I have discovered two general techniques for converting many of the swing-hinged dissections to twist-hinged ones, along with many specialized techniques. So get ready to raise your voices in exclamation!

My techniques for converting swing-hinged dissections require that the dissection have binary hinges and that the hinged pieces be hinge-snug. Then we can replace each swing hinge with a new piece and two twist hinges. The new piece will be the union of two isosceles triangles, one carved from each piece attached to the swing hinge.

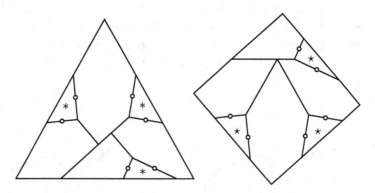

22.7: Twist-hinged dissection of a triangle to a square

I illustrate the technique by converting our old friend, the swing-hingeable dissection of a triangle to a square (Figure 1.1). Figure 22.9 on the left shows the cuts in the triangle, plus dashed edges to indicate the bases of isosceles triangles adjacent to each hinge point. In the twist-hinged dissection (Figure 22.7) I have merged

230

these isosceles triangles together, giving a 7-piece twist-hingeable dissection. Each of the new pieces is a right triangle, which is the case whenever a swing hinge attaches two pieces at corners whose angles sum to 180°.

We see two intermediate configurations in Figure 22.8. On the left, we see the lower left corner of the triangle flipped up, using a pair of twists. Then on the right, we see the lower right corner of the triangle similarly flipped up. Flipping the right corner of what results will then give us the square. If you get a chance to handle a twist-hinged model, you'll find yourself shouting: — *Wow and double-wow!*

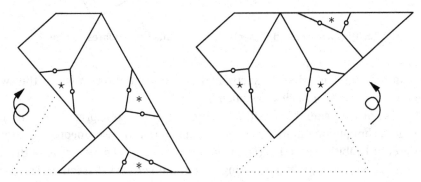

22.8: Intermediate configurations for a twist-hinged triangle to a square

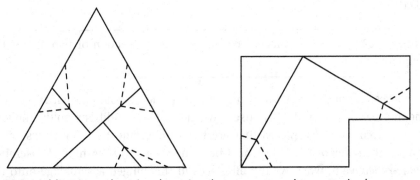

22.9: Adding isosceles triangles: triangle to square and two attached squares

Let's apply my technique to the earliest known hinged dissection, of two attached squares to one (Figure 1.5), using the hinging in Figure 1.6. Figure 22.9 on the right shows the cuts in the two attached squares along with the dashed edges, which indicate the bases of the isosceles triangles. The resulting twist-hinged 5-piece dissection takes shape in Figure 22.10. This time, each hinge attaches two pieces at angles whose sum is 90°. The isosceles triangles are thus both acute triangles, and each new piece has an angle of 135°. It is elementary to show that a new

231

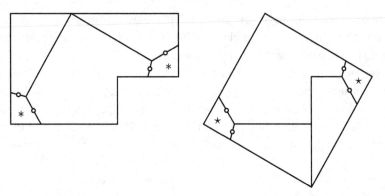

22.10: Twist-hinged dissection of two attached squares to one

piece will have an angle that is exactly half of the angle through which the swing hinge turns. — *Time for verbal vehemence!*

Because of the venerable position accorded to the Pythagorean theorem, let's apply the technique one more time – to the dissection of two unequal squares to one. Careful: Thābit's dissection (Figure 4.4) runs afoul of the requirement that any two pieces connected by a hinge be hinge-snug. But Perigal's dissection (Figure 4.1) satisfies this condition whenever the squares are unequal. When the two squares are equal, we can apply my technique to the dissection in Figure 1.2, giving a 6-piece dissection.

Puzzle 22.1 Describe an 8-piece twist-hinged dissection of two unequal squares to one.

My second technique converts a swing hinge to a single twist hinge with no increase in the number of pieces. We can use it whenever two conditions hold: The swing hinge must connect two pieces that are hinge-snug, and the hinged assemblage on one of the sides of the hinge must be "hinge-reflective." A hinged assemblage is *hinge-reflective* if, when we flip all pieces in this hinged assemblage onto their other side, the hinged assemblage is essentially unchanged. In this case, we glue an isosceles triangle from the piece that is not in the hinged assemblage to the adjacent piece that *is* in the hinged assemblage. The twist hinge must be at the midpoint of the base of that isosceles triangle.

Let's apply this technique to the swing-hinged dissection of a hexagram to two hexagons (Figure 3.25). Pieces B and C are both symmetric and thus constitute hinge-reflective assemblages. We see a hexagon from this dissection in Figure 22.11, with dotted lines indicating the bases of the isosceles triangles; the converted dissection materializes in Figure 22.12. This dissection, when combined with

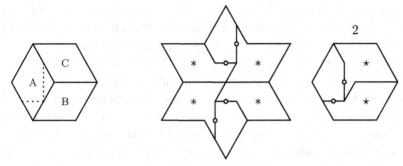

22.11: Add isosceles   **22.12:** Twist-hinged hexagram to two hexagons

the twist-hingeable part of Figure 21.7, gives a completely twist-hingeable dissection of five hexagons to one in 18 pieces. — *Wowie!* Furthermore, this last dissection plus two uncut hexagons gives an 8-piece twist-hingeable dissection of four hexagons to one. — *Yikes!*

Using these two techniques, we can convert many swing-hinged dissections to twist-hinged ones. A notable example is the 8-piece swing-hinged dissection of a pentagon to a square that we can infer from Figure 11.17. Since two of the pieces are hinge-reflective, we replace two swing hinges with twist hinges using the second technique. We then convert the remaining five hinges using the first technique, at the expense of five more pieces. Thus there is a 13-piece twist-hingeable dissection. We cannot convert the 7-piece swing-hingeable dissection in Figure 11.21 because it is not hinge-snug. By noting which dissections in this book are hinge-snug, I have effectively cataloged those dissections that can be so converted.

Let's explore another general technique, this time one that does not involve the conversion of swing hinges. It transforms one parallelogram to another with the same angles, as shown in Figure 22.13. In analogy to the Q-swing and T-swing (which we have seen already in Chapter 3), I have called this the *parallelogram twist*, or *P-twist.* It is reminiscent of a 4-piece dissection discovered by Henry Perigal (1875) that does not turn over any pieces (and thus cannot be twist-hingeable). When the grain runs in one of two directions, my twist-hingeable dissection is grain-preserving.

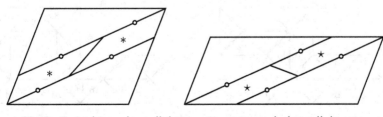

22.13: Twist-hinged parallelogram to same-angled parallelogram

A beautiful feature of the P-twist is that the pieces are cyclicly hinged – our first example of cyclic hinging with respect to twist hinges. The range of achievable dimensions depends on the parallelogram's dimensions a and $b \le a$ and its nonacute angle θ. We can more than double the length, going from a up to (but not including) $a + \sqrt{a^2 + b^2 - 2ab\cos\theta}$. The second term in the preceding expression represents the length of the longer diagonal in the parallelogram and derives from the law of cosines. Since rectangles are parallelograms, the P-twist will transform one rectangle to another. Imagine a coffee table made with a P-twist. — *You won't need caffeine!*

A twist-hinged model does not present so much of a challenge if it already forms one of the two figures, because to form the other figure you need only twist each hinge. On the other hand, if the assemblage forms three or more figures then transforming one of the figures to another is not so simple, since you need not twist all of the hinges. A nifty example is the twist-hinged set of pieces (Figure 22.14) that form any of the twelve pentominoes. For a given n, David Eppstein posed the problem of finding a set of swing-hinged pieces that would form any of the n-ominoes, and computer scientists Erik Demaine, Martin Demaine, David Eppstein, and Erich

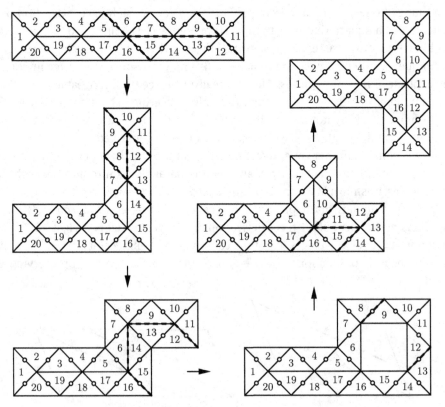

22.14: Twist-hinged pentomino to pentomino

Friedman (1999) gave a solution in $2n - 2$ pieces (see Turnabout 4). So now let's consider the analogous problem for twist hinges.

The twist-hinged pieces in Figure 22.14 are a set of twenty cyclicly hinged isosceles right triangles. A linearly hinged set of the triangles would also work, but the cyclic hinging presents more of a challenge in converting from one pentomino to another. The dashed edges demarcate sections that we can hold rigid as we convert one pentomino to the next in the sequence. Actually, we can use fewer pieces, because we can form each of the pentominoes with triangles 18, 19, 20, 1, 2, and 3 in the same relative position. The same idea holds for larger values of n, so that for any $n \geq 5$, a set of $4n - 5$ twist-hinged pieces will form any of the n-ominoes.

The linearly hinged set of 20 pieces is the two-dimensional analogue of Erno Rubik's snake puzzle, awarded a U.S. patent in (1983). This puzzle consists of 24 identical pieces, each of which is a half of a cube cut along a diagonal plane. Small bolts and nuts connect the pieces into a chain analogous to that in Figure 22.14. So twist hinges have been around since at least the early 1980s! Because the magic snake is three-dimensional, the pieces obstruct each other if you try to perform maneuvers like those in Figure 22.14. However, paper-thin pieces will allow the movement.

In analogy to the previous multi-dissection, for any integer $n > 3$ there exist twist-hinged sets of pieces that allow you to form any n-iamond. This is a twist-hinged version of similar problems for swing hinges, which Demaine et al. (1999) studied. A surprising feature of these sets is that when n is even, you can also form any $(n/2)$-hex from them.

The set consists of $6n$ pieces that form a twist-hinged chain. Half of the pieces are isosceles right triangles, and the other half are kites (symmetrical quadrilaterals). As an example, for $n = 6$, one of the hexiamonds appears on the left in Figure 22.15 along with a tri-hex on the right. We can form each figure by snaking

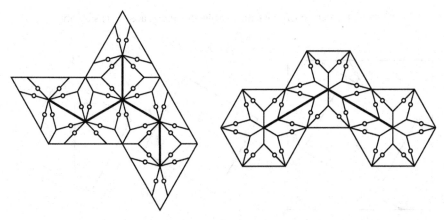

22.15: Twist-hinged hexiamond to tri-hex

the assemblage along the outside boundary, staying outside of a "spine" that spans between the centers of the six triangles of the hexiamond, and similarly for the three hexagons in the tri-hex. I do not know whether we can transform one figure to another if the pieces are cyclicly rather than linearly twist-hinged. I originally drew the accompanying figure to illustrate cyclic twist hinging, and I leave it in that form in the hope of encouraging readers to determine if there are suitable trans-formations in the cyclic case. — *Let's have a clamor of approval!*

Returning to squares, we try three equal squares to one. Monsieur de Coatpont (1877) found a 7-piece dissection in which he turned two pieces over. His dissection is rather pretty, with two of the three squares each having a cut that goes through its center. The solid and dashed edges in Figure 22.16 indicate de Coatpont's dis-section, which we can hinge using swing and flip hinges. With a little more work, we can get a dissection that has only twist hinges. We identify the isosceles trian-gles whose bases are indicated by the dotted lines. Then we add and subtract these

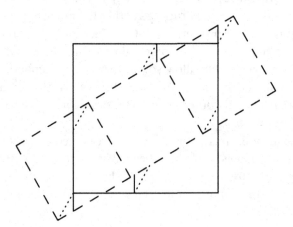

22.16: Derivation of twist-hingeable three equal squares to one

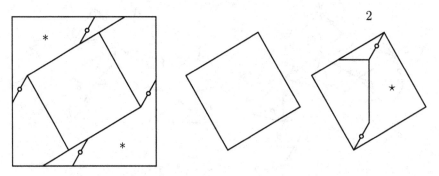

22.17: Twist-hinged dissection of three equal squares to one

isosceles triangles from the pieces to get the 7-piece twist-hinged dissection in Figure 22.17. Note that it is grain-preserving. — *Le pandemonium!*

In Figure 4.25, I gave a 7-piece swing-hinged dissection of squares for $x^2 + y^2 + z^2 = w^2$ when $x + y = w$. When $1 < y/x < 2(1 + \sqrt{2})$, we can produce the twist-hinged dissection in Figure 22.18. We cut the y-square into two pieces and then arrange the x-square and the twist of the y-square to fill out all but a rectangle in the w-square. Applying a P-twist to the z-square gives the rectangle that completes the dissection.

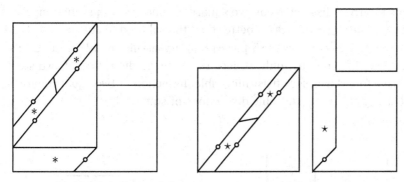

22.18: Twist-hinged three squares to one with $x + y = w$

Alfred Varsady (1989) found a 9-piece unhingeable dissection of five equal triangles to one that turns over two pieces and thus seems a natural for a conversion into twist-hingeable form. The basis for Varsady's dissection is shown in Figure 22.19, with the five equal triangles in solid edges and the large triangle in dashed edges. To obtain a twist-hingeable dissection, we add isosceles triangles as indicated by

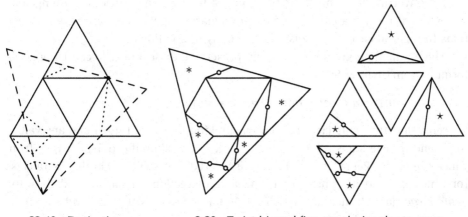

22.19: Derivation   **2.20:** Twist-hinged five equal triangles to one

237

dotted edges near the right and near the top. We first take the left portions of the two triangles on the lower left and swing them around the bottom vertex. We then convert this swing hinge to twist hinges by using the isosceles triangles indicated by the dotted edges on the bottom. However, we cannot swing one of these pieces around because it is not attached, so we make space for it in the bottom triangle and swing around the piece that it replaces. The boundaries for these two pieces are also dotted. Adding and subtracting these triangles from the various pieces gives the 11-piece twist-hinged dissection for us to ponder in Figure 22.20. — *Let out a yell!*

Let's return to dissections of two squares to one. As with swing-hinges, special approaches handle certain cases better than the solution to Puzzle 22.1. For example, for $3^2 + 4^2 = 5^2$ we have the 5-piece grain-preserving twist-hingeable dissection in Figure 22.21. Is this the only 5-piece twist-hinged dissection of two squares to one? I have found a 6-piece twist-hingeable dissection of two equal squares to one as well as 7-piece twist-hingeable dissections of squares for $5^2 + 12^2 = 13^2$ and for $8^2 + 15^2 = 17^2$.

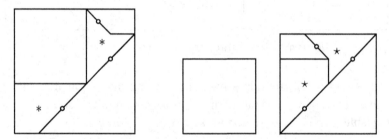

22.21: Twist-hinged dissection of squares for $3^2 + 4^2 = 5^2$

By converting swing hinges to twist hinges, we can find a 9-piece twist-hingeable dissection of two squares to two different squares. Again, there are better dissections for special cases. There is a large family of solutions to $x^2 + y^2 = z^2 + w^2$ for which a method gives 6-piece grain-preserving dissections. Recall Fibonacci's formula from Chapter 8:

$$x = mp + nq, \quad y = mq - np, \quad z = mq + np, \quad w = |mp - nq|.$$

We can define the Penta class of solutions, taking $m = 2$ and $n = 1$ and any values for p and q. The sums $x^2 + y^2$ and $z^2 + w^2$ then simplify to $5(p^2 + q^2)$, from which the name "Penta" arises. I have discovered that solutions in the Penta class for which $q > 2p$ have nice twist-hingeable dissections. I call this subclass the *Penta-large* class. It includes $5^2 + 5^2 = 7^2 + 1^2$, $6^2 + 7^2 = 9^2 + 2^2$, and $7^2 + 9^2 =$

$11^2 + 3^2$ (to name a few) in the case $p = 1$. But we need not restrict p and q to be whole numbers. Real numbers work just as well, as long as $q > 2p$. — *And the crowd goes wild!*

Method Penta-L: For squares in the Penta-large class.

1. Cut a y-square from the upper left corner of the z-square.
2. Starting $q - p$ below the upper right corner of the z-square,
   cut a diagonal line to a point $2p$ to the left and $2p$ down.
3. Starting $q - p$ to the right of the lower left corner of z-square,
   cut a diagonal line to a point $2p$ to the right and $2p$ up.
4. Starting p below the upper right corner of the z-square,
   cut a line to a point p to the left,
   and from there to the upper right corner of y- square.

In Figure 22.22 I give the dissection for $7^2 + 6^2 = 9^2 + 2^2$, for which $p = 1$ and $q = 4$. I describe the approach for finding 6-piece twist-hingeable dissections as Method Penta-L. When $q = 2p$, we have $w = 0$, and x, y, and z, are multiples of 4, 3, and 5, respectively. The method works for this case, producing the dissection of Figure 22.21. I have also found an unrelated 6-piece dissection for $7^2 + 4^2 = 8^2 + 1^2$.

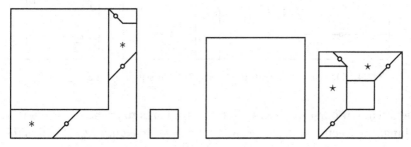

22.22: Twist-hinged dissection of squares for $7^2 + 6^2 = 9^2 + 2^2$

For three unequal squares to one, I have found 6-piece twist-hingeable dissections for $2^2 + 3^2 + 6^2 = 7^2$, $1^2 + 4^2 + 8^2 = 9^2$, and $7^2 + 4^2 + 4^2 = 9^2$. Of course, there is a 4-piece twist-hingeable dissection of squares for $1^2 + 2^2 + 2^2 = 3^2$, but that is too easy to show. More interesting is a twist-hingeable dissection of Greek Crosses for $1^2 + 2^2 + 2^2 = 3^2$. If we convert the swing hinges in Figure 8.9 to twist hinges, we need 14 pieces. In Figure 22.23, I give a symmetrical 13-piece twist-hinged dissection.

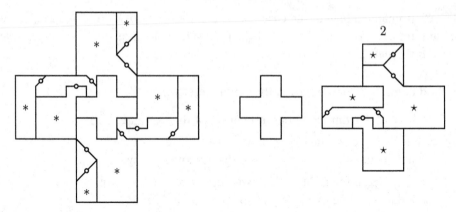

22.23: Twist-hinged dissection of Greek Crosses for $1^2 + 2^2 + 2^2 = 3^2$

What happens if we require that every cut in a dissection of squares be parallel to one of the sides of the square? This restriction seems to make the puzzle rather more challenging. For squares for $3^2 + 4^2 = 5^2$, I have found the 9-piece grain-preserving dissection that maintains its rectitude in Figure 22.24.

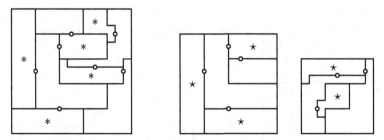

22.24: Oriented twist-hinged dissection of squares for $3^2 + 4^2 = 5^2$

Let's reorient ourselves and switch from squares to triangles. There is a nifty twist-hinged dissection of triangles for $3^2 + 4^2 = 5^2$ in Figure 22.25. Is this the only 4-piece twist-hinged dissection of two triangles to one?

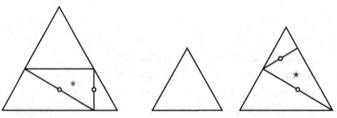

22.25: Twist-hinged dissection of triangles for $3^2 + 4^2 = 5^2$

Having succeeded with triangles, let's try hexagons for $3^2 + 4^2 = 5^2$. I found the 8-piece twist-hinged dissection that stands jewel-like in Figure 22.26. Two intermediate configurations reflect a lovely symmetry in Figure 22.27. As with the previous dissection, the technique is inspired trial and error – with a bit more inspiration in this case. — *A shriek of delight!*

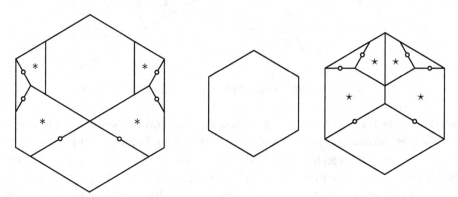

22.26: Twist-hinged dissection of hexagons for $3^2 + 4^2 = 5^2$

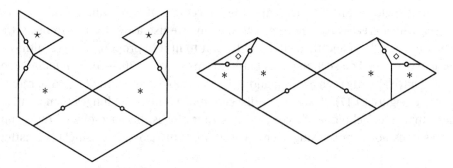

22.27: Intermediate configurations of the 4-hexagon

Given our technique for converting swing hinges to twist hinges, some twist-hingeable dissections are easy. In particular, we can find an 11-piece twist-hingeable dissection of an {8/2} to a square that is based on Figure 21.9. Also, we can find a 13-piece twist-hingeable dissection of an {12/2} to a hexagon that is based on Figure 21.11.

A bit more challenging is the dissection of a hexagram to a triangle, since we have no dissection from the previous chapter to start with. Furthermore, not all of the hinges in Figures 15.13 and 15.14 are hinge-snug; hence we cannot use our swing-to-twist conversions. Instead we cut the hexagram into three equal pieces, twist two,

241

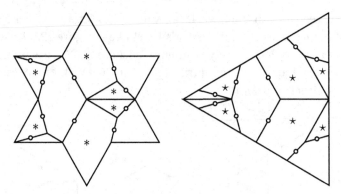

22.28: Twist-hinged dissection of a hexagram to a triangle

and then make four cuts with swing hinges. We then convert those swing hinges to twist hinges, producing the 11-piece dissection that confounds us in Figure 22.28.

To get a twist-hingeable dissection of a dodecagon to a square, we start with the 6-piece unhingeable dissection by Harry Lindgren (1951). The solid lines in Figure 22.29 indicate Lindgren's dissection, and dotted lines indicate isosceles triangles that I switch from one piece to another, as well as additional cuts. We add two isosceles triangles to the equilateral triangle and use two twist hinges to flip the resulting triangle around. Then we use the conversion of two swing hinges to twist hinges, which creates two more pieces. Finally, we use a twist hinge to bring the concave piece along and then slice and twist it to fit it in properly. The resulting 9-piece twist-hinged dissection exalts before us in Figure 22.30. — *What a hullabaloo!*

We have already seen a swing-and-twist-hingeable dissection of a pentagram to a square (Figure 21.17). If we want a dissection with only twist hinges then we need more than the technique for converting swing hinges, because one of the swing hinges does not connect hinge-snug pieces. I cut the isosceles triangle consisting

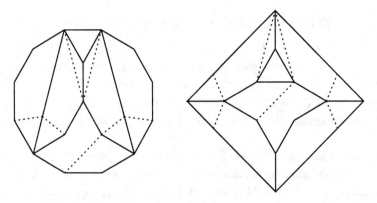

22.29: Derivation of a twist-hingeable dodecagon to a square

242

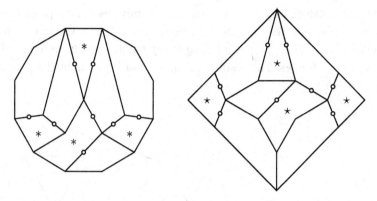

22.30: Twist-hinged dissection of a dodecagon to a square

of two pieces in the lower left point of the pentagram differently and have it twist to the right. Because of the short length of some of the edges, I do not show that particular twist hinge in Figure 22.31. And since the right triangle next to it is also small, I do not show the twist hinges on its legs; nor do I label it with an asterisk in the pentagram and a star in the square. Even with 13 pieces, this dissection is such a lovely surprise. — *Scream at the top of your lungs!*

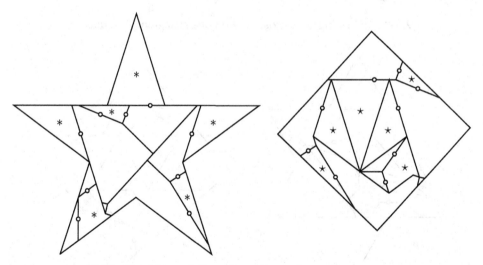

22.31: Twist-hinged dissection of a pentagram to a square

We have seen a swing-hingeable dissection of a hexagram to a square in Figure 11.24, but that dissection is also not hinge-snug. Let's try again, using a T-strip element that we form in the following way. Slice the hexagram in half with a line segment from the midpoint of one side to the midpoint of the opposite side. We

243

then join the two halves with a swing hinge. Next, move two of the points out of the way, using twist hinges. We then have a strip element that we use in the crossposition in Figure 22.32. This gives rise to five swing hinges, which we replace by right triangles and twist hinges. The resulting 13-piece fully twist-hinged dissection is in Figure 22.33. — *Yo! What a surprise!*

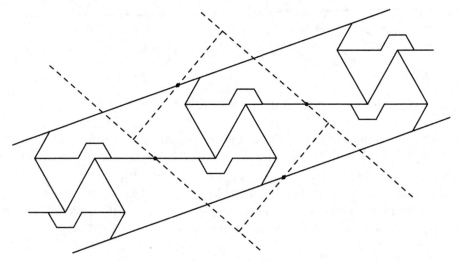

22.32: Twist-hingeable crossposition of a hexagram to a square

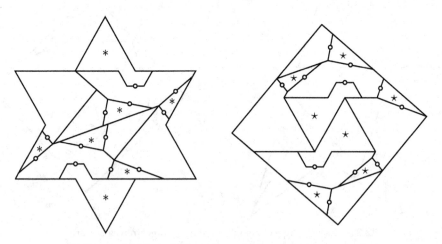

22.33: Twist-hinged dissection of a hexagram to a square

We have seen a swing-hingeable dissection of a Latin Cross to a square in Figure 11.48, but that dissection is not hinge-snug. Let's try again, using the twist-hingeable T-strip element of Figure 22.34. If we crosspose as in Figure 22.35, we

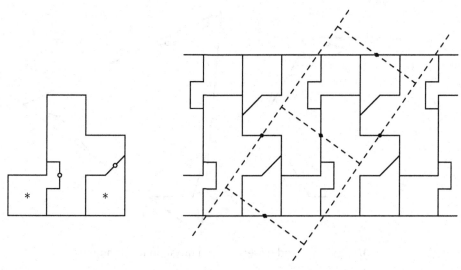

22.34: Strip element   **22.35:** Crossposition of Latin Cross and square

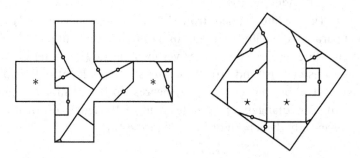

22.36: Twist-hinged dissection of a Latin Cross to a square

create four swing hinges. We replace the swing hinges with right triangles and twist hinges to produce an 11-piece fully twist-hinged dissection (Figure 22.36). Because of the small size of one of the four right triangles, I do not show the twist hinges on it. Also, I do not label the four right triangles with asterisks or stars.

We next explore a family of dissections that is truly extraordinary. For any $p > 2$, there is a $(2p + 1)$-piece twist-hingeable dissection of a $\{2p\}$ to a $\{p\}$. Furthermore, these elegant dissections exhibit p-fold rotational symmetry. One such twist-hinged dissection – of a hexagon to a triangle – startles us with its simplicity and symmetry in Figure 22.37.

The dissection technique is so simple and symmetric that it is hard to believe that it was not already known. Overlay the $\{2p\}$ and the $\{p\}$ so that their centers coincide and every side of the $\{p\}$ bisects a side of the $\{2p\}$. For each vertex A of the $\{p\}$, draw a line segment to the nearest vertex B of the $\{2p\}$. For each such

245

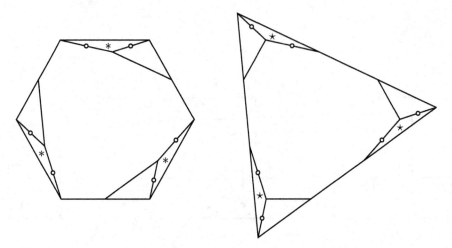

22.37: Twist-hinged dissection of a hexagon to a triangle

vertex B of the $\{2p\}$, identify the vertex C of the $\{2p\}$ such that a side of the $\{p\}$ bisects side BC. Draw a line segment from C to the nearest side of the $\{p\}$, with the line segment parallel to the line segment AB. From the other endpoint D of this line segment, draw a line segment to the next vertex B' of the $\{2p\}$ after C. In conjunction with the sides of $\{2p\}$ and $\{p\}$, these line segments identify the appropriate cuts to make. Let A' be the vertex of $\{p\}$ nearest to B'. By simple trigonometry, using an identity for cotangents of double an angle, the length of the line segment from D to A' is exactly the length of the side of the $\{2p\}$.

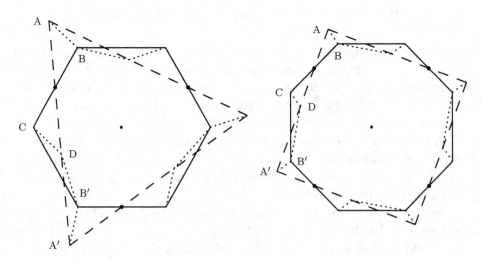

22.38: Overlaying a hexagon and a triangle, and an octagon and a square

Figure 22.38 illustrates the technique with the overlay of a hexagon and a triangle on the left and an octagon and a square on the right. I discovered these dissections by shifting the tessellations used in the "completing the tessellation" technique. The octagon to square comes from directly modifying Figure 12.1 to produce the second superposition in Figure 22.39, and the hexagon to triangle derives from overlaying a tessellation of hexagons with a tessellation of triangles and hexagons. The rest of the dissections do not derive directly from tessellations, since there is (for example) no tessellation of decagons with any other regular figure. Yet the debt to the completing the tessellation technique is obvious, so that I call this method "completing the pseudotessellation." — *Oh, I get by with a little help from my pseudofriends!*

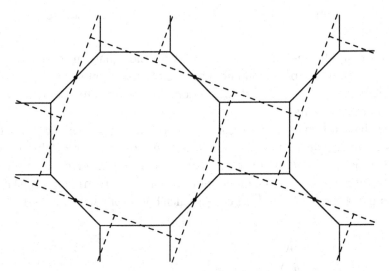

22.39: Second superposition of tessellations for an octagon and a square

The corresponding 9-piece twist-hinged dissection of an octagon to a square lords over us in Figure 22.40. The next two examples in this family are of a decagon to a pentagon and a dodecagon to a hexagon. Harry Lindgren (1964b) found a 10-piece unhingeable dissection for the former, and I found a 9-piece unhingeable dissection for the latter. None of these previous dissections even hint at the possibility of this lovely twist-hingeable family. — *Deafening tumult!*

An unexpected dividend is that, if we turn over the square in Figure 22.40 and overlay it on the square in Figure 1.1, none of the lines cross! Applying the same technique as in Figure 22.7 to convert the swing hinges to twist hinges, we get a 15-piece 3-way twist-hingeable dissection of an octagon to a square to a triangle. In this case we need to have smaller right triangles replacing the swing hinges. We

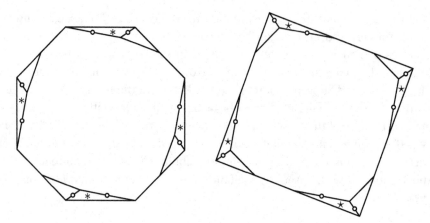

22.40: Twist-hinged dissection of an octagon to a square

can also overlay the square of Figure 22.40 with the square in the 13-piece twist-hinged dissection of a pentagon and a square discussed on page 233. The result is a 21-piece 3-way twist-hingeable dissection of an octagon to a square to a pentagon. — *Raucous bedlam!*

After the last family of dissections, the surprise is that there are more dissections of a similar nature. Consider any $p > 4$ and $q < p/2$ for which the distance from the center of a $\{p\}$ to a vertex is greater than the distance from the center of a $\{p/q\}$ of equal area to the midpoint of a side. That restriction translates into the condition $4\cos\frac{(q-1)\pi}{p}\cos\frac{q\pi}{p} \geq \cos\frac{\pi}{p}$, which holds for whole numbers whenever

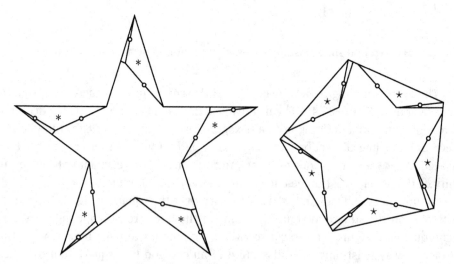

22.41: Twist-hinged dissection of a pentagram to a pentagon

$p \geq 3q - 1$. Then there is a $(2p + 1)$-piece twist-hingeable dissection of the $\{p/q\}$ to the $\{p\}$. The dissection exhibits p-fold rotational symmetry, as in the 11-piece twist-hinged dissection of a pentagram to a pentagon that almost preens in Figure 22.41. This compares with a 9-piece unhingeable dissection by Gavin Theobald.

As in the previous family of dissections, we produce these dissections by overlaying the two figures so that their centers coincide and each side of the $\{p\}$ bisects a side of the other figure. The overlay of pentagram and pentagon is on the left in Figure 22.42, and the overlay of a $\{7/2\}$ and a heptagon is on the right. Again, it is appropriate to call this the "completing the pseudotessellation" technique. What a marvelously rich technique! For each pair of figures $\{p/q\}$ and $\{p\}$, there is a second $(2p + 1)$-piece twist-hingeable dissection. This follows because, if the $\{p\}$ can bisect a side of the $\{p/q\}$, then there are two different ways to do the bisection. The second way leads to a second dissection of pentagram and pentagon in which the slivers are even thinner.

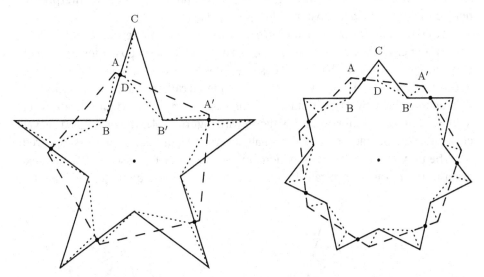

22.42: Overlaying a pentagram and a pentagon, and a $\{7/2\}$ and a heptagon

Another example is the twist-hinged dissection of a $\{7/2\}$ to a heptagon that sneaks up on us in Figure 22.43. I do not know of any previous dissection of a $\{7/2\}$ to a heptagon, hingeable or otherwise. We can't produce a dissection of a $\{7/3\}$ to a heptagon by this method, because $7 < 3*3-1$. However, I have a 17-piece dissection of an $\{8/2\}$ to an octagon. Can you find it? It compares with a 24-piece unhingeable approximate dissection from the anonymous circa 1300 *Interlocks* manuscript.

Thus we seem to have two remarkable families of dissections, of $\{2p\}$ to $\{p\}$ and of $\{p/q\}$ to $\{p\}$, both using essentially the same dissection technique. Is it

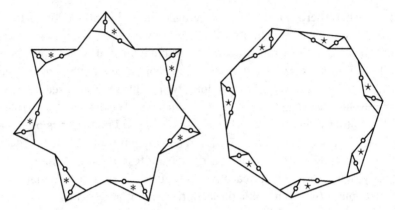

22.43: Twist-hinged dissection of a {7/2} to a heptagon

possible that we have one large family instead of two? Indeed, we do, in a fashion similar to what we encountered at the end of Chapter 9. Thus we interpret the polygon {2p} as the pseudostar {p/0.5}. — *A thunderous roar!*

We recall that, for pseudostar {p/q}, q need not be an integer; we only require a real number in the range $1/2 \le q < p/2$. The dissection of a pseudostar {p/q} to polygon {p} will work whenever $4 \cos \frac{(q-1)\pi}{p} \cos \frac{q\pi}{p} \ge \cos \frac{\pi}{p}$. When equality holds in this relation, the dissection has only $p + 1$ pieces rather than $2p + 1$; in this case, let's denote the q by $\lambda(p)$. I illustrate this case for $p = 5$ in Figure 22.44, where $\lambda(5) \approx 2.0813935$. Comparing this figure with Figure 22.41, the pseudostar is very close to a true pentagram, and the small slivers in Figure 22.41 have disappeared.

The previous family of dissections includes a 13-piece dissection of the hexagram to the hexagon. We can do better. I found a 7-piece unhingeable dissection

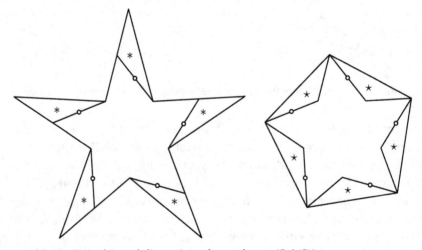

22.44: Twist-hinged dissection of pseudostar {5/λ(5)} to a pentagon

250

of a hexagram to a hexagon, which is not the best possible. Not only does the dissection use one piece more than the best known, but it has three pieces that we must turn over. However, the pieces that we turn over are the key to a nifty twist-hingeable dissection. The solid lines in Figure 22.45 indicate the 7-piece un-hingeable dissection. We add two isosceles triangles (indicated by dotted lines) to each of the three small triangles, giving three triangles that we can twist-hinge.

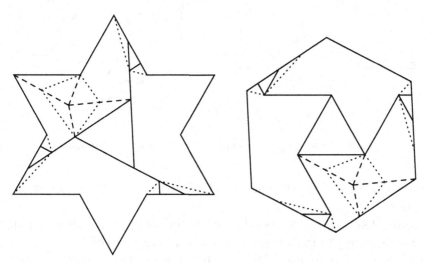

22.45: Derivation of a twist-hingeable hexagram to a hexagon

Producing the three new triangles does not yield a dissection that is completely twist-hingeable, because there is an equilateral triangle that must remain in the center as other pieces twist around. (The original dissection is the result of completing the tessellation, using equilateral triangles together with hexagrams and also with hexagons.) Following the lead in Chapter 12, we identify two irregular triangles that can swap positions, as shown with dashed edges. To make these new pieces twist-hingeable, we introduce more isosceles triangles (dotted edges). As luck would have it, we can glue all four of these isosceles triangles together, producing a trapezoid for our 10-piece twist-hinged dissection in Figure 22.46.

An observant reader will see a cyclic hinging of eight of the pieces, and a skeptical reader may wonder if this actually works. I was not sure whether something so remarkable was possible until I had constructed and tested a rough model out of thin foamboard and toothpicks. Then I verified mathematically that it does indeed work. There are five pieces centered on the trapezoid that play somewhat the same role as the large pieces. To convert the hexagram to hexagon, flip over the large pieces and the trapezoid-centered five, rotating them simultaneously about the axes shown as dotted lines, while partially turning the small

251

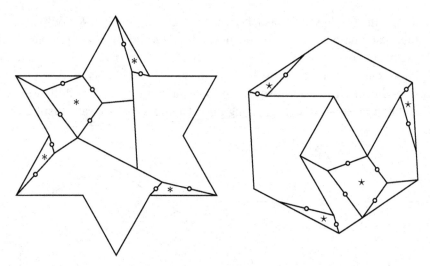

22.46: Twist-hinged dissection of a hexagram to a hexagon

triangles so as to accommodate the differing levels of the twist hinges on the parallel edges.

Figure 22.47 shows a perspective view of the configuration after rotating the large pieces by 90° from their position in the hexagram. The three dotted lines identify an imaginary equilateral triangle that stays fixed as the pieces rotate. Each vertex of this triangle is the center of a smaller equilateral triangle (not shown) adjacent to the long edge of a small triangle. Furthermore, the axis of the twist hinge

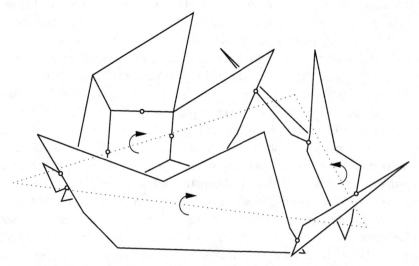

22.47: Intermediate configuration of a hexagram to a hexagon

between a large piece and a small triangle pierces the center of that smaller equilateral triangle. As the large pieces complete their turning, the small pieces return to their original side up. — *Now we can really whoop it up!*

This last dissection suggests the possibility of an improved twist-hingeable dissection for a dodecagon to a hexagon. The dissection in Figure 22.45 derives from a superposition of two tessellations produced by the "completing the tessellation" technique. We have already superposed two such tessellations for dodecagons and hexagons in Figure 12.5. Let's superpose these two tessellations again, this time positioning a small triangle in the center of a dodecagon and another in a hexagon (Figure 22.48).

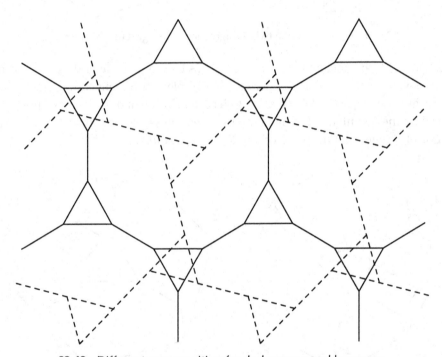

22.48: Different superposition for dodecagons and hexagons

This different superposition produces a 7-piece dissection (solid lines in Figure 22.49) that is rather similar to the 7-piece dissection of hexagram to hexagon in Figure 22.45. Again, we have three small triangles that we must turn over. We again add two isosceles triangles (dotted lines) to each of the small triangles, producing three triangles that we can twist-hinge.

Once again we must transfer an equilateral triangle from the center of one figure to the center of the other. We can use the same trick from Chapter 12 as adapted in our hexagram-to-hexagon dissections. The only difference is that we now make

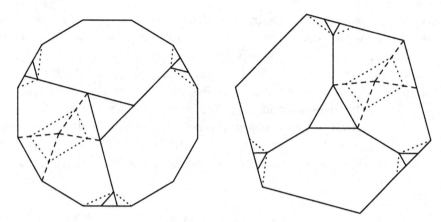

22.49: Derivation of a twist-hingeable dodecagon to a hexagon

the trapezoid a bit smaller than the largest possible, so that one of the other pieces is not so small. As a result, we derive the a 10-piece twist-hinged dissection that surely glows in Figure 22.50. This is a 3-piece improvement over the dissection produced by the technique of completing the pseudotessellation. Again, we are able to cyclicly hinge eight of the pieces. — *Boisterous exuberance!*

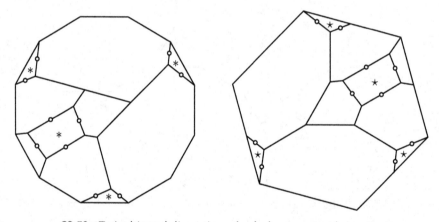

22.50: Twist-hinged dissection of a dodecagon to a hexagon

It seems to be easy to find twist-hingeable dissections of four equal figures to one. To find a twist-hinged dissection of four pentagons to one (Figure 22.51), we start with a 6-piece unhingeable dissection by Harry Lindgren (1964b). Lindgren cut 72°-rhombuses out of two of the pentagons and left the other two pentagons uncut. We can follow the same plan here, but we need an extra cut in each of the cut pentagons to turn the pieces around appropriately.

254

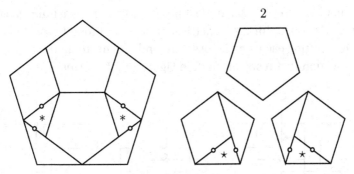

22.51: Twist-hinged four pentagons to one

A *notched rectangle* is a rectangle that has a smaller rectangle cut from one of its corners. Lindgren (1964b) gave a 6-piece unhingeable dissection of a notched rectangle to four smaller congruent ones, and Robert Reid found two 5-piece dissections in which he turned over two pieces.

Puzzle 22.2 Find a 7-piece grain-preserving twist-hingeable dissection of four notched rectangles to one.

There is a pretty 12-piece twist-hinged dissection of four octagons to one, basking in Figure 22.52. We cut and hinge all four small octagons in the same way, producing a large octagon with 4-fold rotational symmetry. Since we can use the same number of pieces for four-to-one dissections of pentagons (whether swing-hingeable or twist-hingeable) and similarly for octagons, does this indicate a pattern? The pattern does not seem to apply to hexagons, but what good rule does not have an exception?

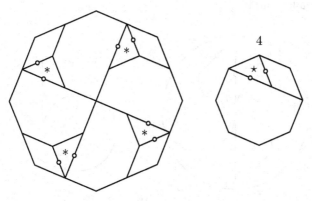

22.52: Twist-hinged four octagons to one

The Latin Cross presents an interesting challenge for a twist-hingeable four-to-one dissection. Looking for a symmetrical dissection, I found nothing better than 14 pieces. But cutting each small cross differently seems to do the trick, giving the 12-piece dissection that tests our faith in Figure 22.53. — *Yippee!*

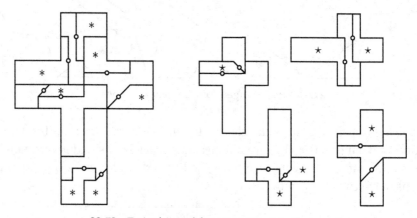

22.53: Twist-hinged four Latin Crosses to one

And now it's time for our last hurrah! Let's come "full circle" – back to a twist-hinged dissection of a curved figure, like William Esser's dissection of an oval to a heart in Figure 1.9. One of the more charming of the unhingeable curved dissections is Sam Loyd's dissection of a circular disk to two oval seat tops in just four pieces, described by Henry Dudeney (*Strand,* 1927a). Those figures seem resistant to a swing-hingeable dissection, but I have found the 6-piece twist-hinged dissection that curves conclusively in Figure 22.54. If we rotate the handholds in the middle of each seat top by 90°, then a 10-piece twist-hingeable dissection is possible. Can the reader find it?

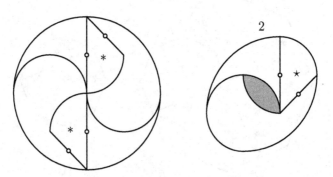

22.54: Twist-hinged disk to two oval seat tops

— Hey!
We've gone from swinging to twisting to shouting,
With a big dose of moving and grooving.
So whenever you get that yearning,
Just remember ... to keep on turning!

CHAPTER 23

Puzzles Unhinged

This chapter title may seem a bit puzzling. When we say that something becomes unhinged, we mean that it becomes unstable, unsettled, or disrupted. Thus, when a puzzle becomes unhinged, it must be because someone has supplied the solution. So, puzzles beware; here are those solutions!

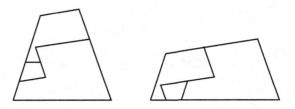

Solution 3.1. Another 3-piece Q-step dissection for quadrilaterals

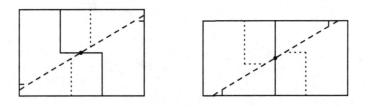

Solution 3.2. Another hingeable dissection of a gnomon to a square

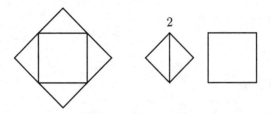

Solution 4.1. Hingeable dissection of squares for $x = y = z/\sqrt{2}$

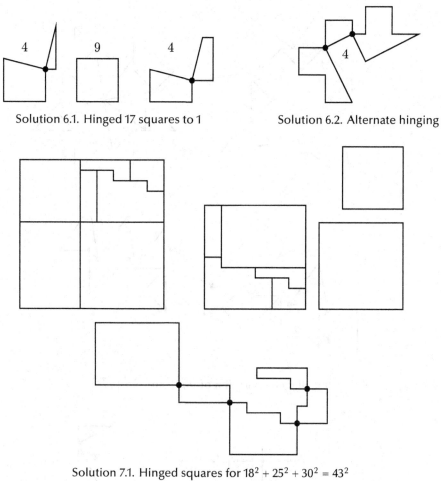

Solution 6.1. Hinged 17 squares to 1

Solution 6.2. Alternate hinging

Solution 7.1. Hinged squares for $18^2 + 25^2 + 30^2 = 43^2$

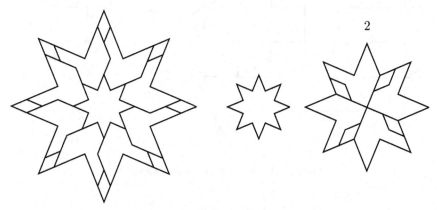

Solution 8.1. Hingeable dissection of {8/3}s for $1^2 + 2^2 + 2^2 = 3^2$

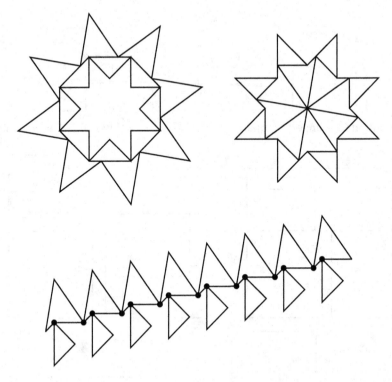

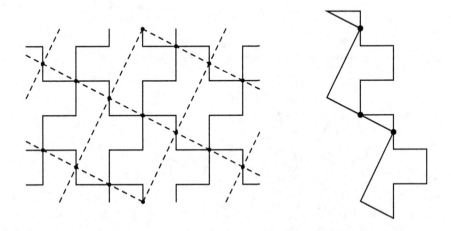

Solution 9.1. Hinged dissection of {8/3}s for $1^2 + (\sqrt{2})^2 = (\sqrt{3})^2$

Solution 10.1. For a Greek Cross to a square

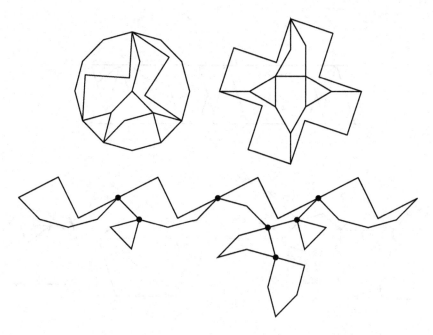

Solution 10.2. Hinged dissection of a dodecagon to a Greek Cross

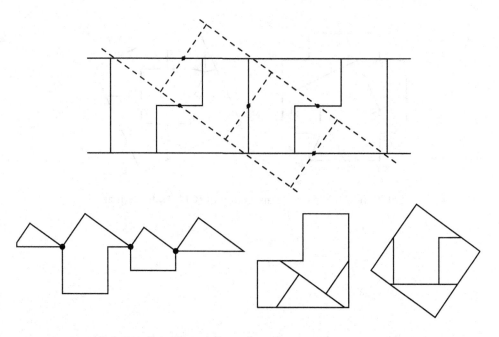

Solution 11.1. Hinged dissection of a gnomon to a square

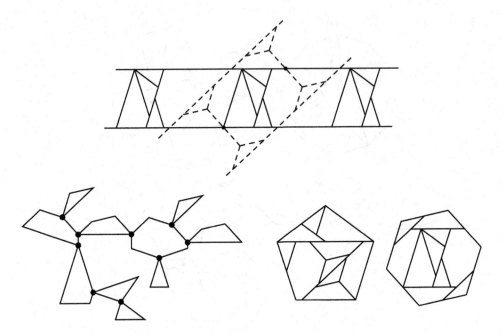

Solution 11.2. Hinged dissection of a hexagon to a pentagon

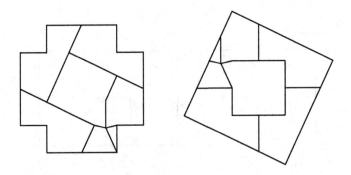

Solution 12.1. Hingeable dissection of {G(4/7)} to a square

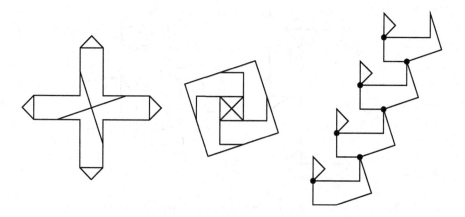

Solution 13.1. Hinged dissection of $\hat{M}_1$ to a square

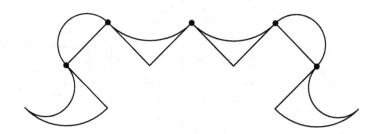

Solution 14.1. Hinged pieces for Leonardo's motif to a rectangle

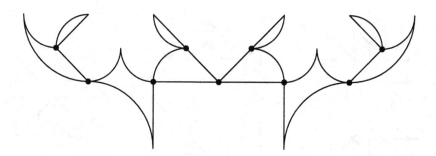

Solution 14.2. A different hinging for Leonardo's motif to a square

Solution 14.3. Hingeable Globular Latin Cross to a Greek Cross

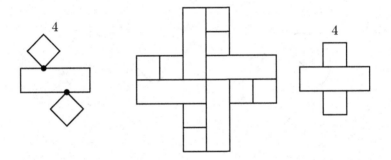

Solution 16.1. Hinged dissection of four Greek Crosses to one

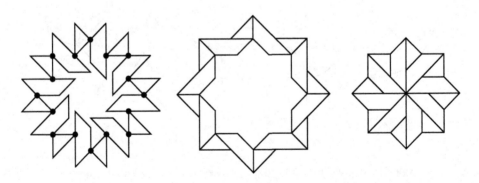

Solution 17.1. Hinged dissection of two {8/2}s to one

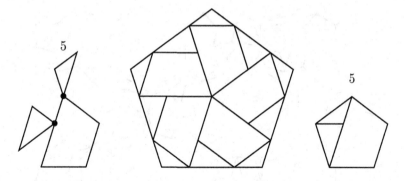

Solution 17.2. Hinged dissection of five pentagons to one

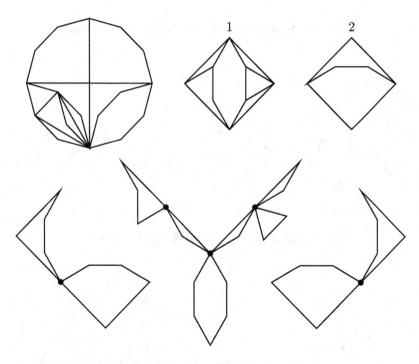

Solution 18.1. Hinged dissection of a dodecagon to three squares

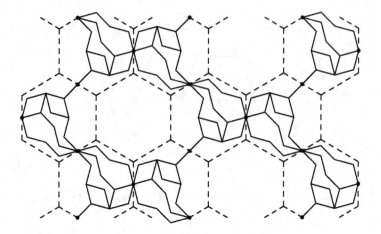

Solution 18.2. Superposition for one 12/2 to three hexagons

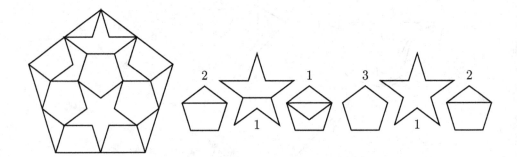

Solution 18.3. Eight {5}s and two {5/2}s to a large {5}

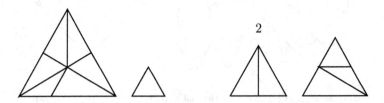

Solution 18.4. Triangles for $3^2 + 1^2 = (\sqrt{3})^2 + (\sqrt{3})^2 + 2^2$

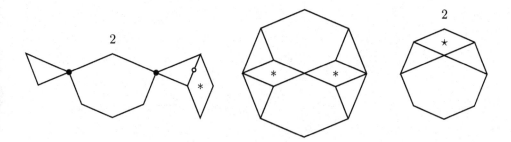

Solution 21.1. Swing-and-twist-hinged dissection of two octagons to one

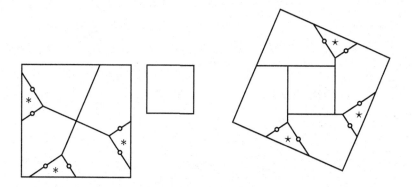

Solution 22.1. Twist-hinged dissection of two unequal squares to one

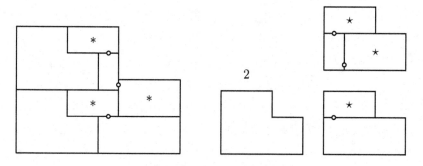

Solution 22.2. Twist-hinged four notched rectangles to one

Afterword

For a long time, I had tentatively entitled this book as *Dissections Too! Swingin'*. Early references to it may still use this title.

Irving Mills wrote the lyrics and Duke Ellington composed the music for "It Don't Mean A Thing If It Ain't Got That Swing." Irving Mills was a musician, talent manager, publisher, composer, and lyricist, and Duke Ellington was a renowned jazz musician, band leader, and composer. A recently available source of information about Henry Dudeney is an excerpted version of the diaries of his wife, Mrs. Henry Dudeney (1998). Throughout this book, I have tried to avoid the repetitious use of the phrase "in my first book." Readers wishing to track down dissections for which I have given no specific reference should check there first.

The 4-piece hinged dissection of a triangle to a square has been described in books by H. Martyn Cundy and A. P. Rollett (1952), Martin Gardner (1961), Howard Eves (1963), Harry Lindgren (1964b), Hugo Steinhaus (1969), Isaac Schoenberg (1982), Ian Stewart (1987), David Wells (1991), myself (1997), Eric Weisstein (1998), Donald Benson (1999), and Jin Akiyama and Gisaku Nakamura (2000c). As inspired as it is, the hinged dissection also seems to be a bit jinxed. The third edition of Steinhaus's book (1969) claimed that dimensions along the base of the triangle were in the ratio of $1:2:1$. This is clearly wrong, as Don Crowe and Isaac Schoenberg (1984) pointed out. Ian Stewart (1987) and Eric Weisstein (1998) placed a hinge at the wrong spot when they illustrated the hinged assemblage. Katharine McKenna (1988) patented a geometric toy that was essentially the hinged dissection, only 86 years after its discovery!

David Eppstein (2001) has created a method that produces a hinged dissection of any asymmetric polygon to its mirror image. The method identifies a chain of kite-shaped pieces by using a circle-packing algorithm to generate a quadrilateral mesh.

If a dissection is swing-hingeable, then it would seem that it cannot be translational because we are using rotation to move the pieces. However, there are dissections that are both translational and swing-hingeable: either a piece that rotates is rotationally symmetrical, or several pieces of the same shape move to different

269

positions depending on the property being demonstrated. The dissections in Figures 1.2 and 1.3 are also translational.

The hinges that I have considered use the simplest of the lower pairs of linkages – namely, the revolute joint (or pin joint), which permits only relative rotation. For further background, see the book by Kramer (1992). Gantes, Giakoumakis, and Vousvounis (1997) and Wasfy and Noor (1998) discussed the somewhat similar "deployable structures," which have applications in outer space. The fascination that I and others have felt toward rotating hinged dissections may have a neurological basis. Pinker (1997) reported on experiments – performed by Cooper and Shepard (1973) and followed up on by Tarr and Pinker (1989) and Tarr (1995) – that examined the way in which our visual systems perform mental rotation of objects.

Instead of William Esser's ellipsoid, Ernst Lurker (1985) used the shape of a pill, so that the resulting heart is of a slightly different shape. Lurker's design first appeared in (1984) as a 7-inch nickel-plated aluminum model in a limited edition of 80 that was distributed by Bayer in Germany to promote a heart medicine. In the same year, Bayer and Pfizer also produced smaller models made from red plastic. The similar designs of Esser and Lurker seem to be coincidental; Esser filed for a patent in 1983, and Lurker produced an unpublished model in 1981. The same idea has been used in a key-chain light that turns on when twisted one direction.

Alpay Özdural is preparing a book that contains a translation and analysis of the *Interlocks* manuscript. Cantor (1907) and Allman (1889) discussed the dissection of two equal squares to one as it appears in Plato's works.

In my first book, I classified some dissections as being "partially hinged." By partially hinged I meant that the pieces are not all connected in one hinged assemblage. Unfortunately, I made no distinction between a dissection of one figure to one figure, such as a hexagon to a triangle, and a dissection of several figures to one, such as three squares to one. In this book, I have discarded the notion of partial hinging.

Readers with a substantial background in mathematics may find it interesting to compare the mathematical model for dissection in Chapter 2 with treatments by Sah (1979), Wagon (1985), and Laczkovich (1990).

I give a catalog of both animated and unanimated dissections on the world wide web at ⟨http://www.cs.purdue.edu/homes/gnf/webdiss.html⟩.

We can derive formulas for the Penta-Penta-plus class as follows. First substitute $m = 2$ and $n = 1$ into Fibonacci's formula, producing $x = 2p + q$, $y = 2q - p$, $z = 2q + p$, and $w = 2p - q$. Then substitute into these, using the last y-value $2q - p$ in place of p and the last x-value $2p + q$ in place of q.

Woepcke (1855) and Fourrey (1907) each described Abū'l-Wafā's dissections of squares. Allman (1889) discussed the dissection of three equal triangles to one as it appeared in Plato.

270

Dickson (1920, pp. 225–6) gave a more complete discussion of the history of integer solutions for $x^2 + y^2 = z^2 + w^2$. In my first book, I defined the Pythagoras-extended class of solutions to $x^2 + y^2 = z^2 + w^2$ to apply whenever $m = n + 1$ or $m = n + 2$ in Fibonacci's formula. Since $m = n + 1$ is the restriction that gives Pythagoras's class in Diophantus's formula for $x^2 + y^2 = z^2$, I gave this class the name the *Pythagoras-extended* class. The Plato-extended class, which corresponds to the subclass when $n = 1$, is also equivalent to the subclass when $m = n + 2$. So perhaps I should have defined the Pythagoras-extended class to apply only for $m = n + 1$.

Banks (1999) discussed variations in pentagrams in "Skinny Pentagrams and Fat Pentagrams" and "Some Geometrical Features of Pentagrams." He represented the "puffed-out" stars somewhat differently, but he still identified the limiting case of a fat pentagram as being a decagon. Synder (1993) discussed several star projections of the Earth. As discussed by Pattison (1960), the Association of American Geographers adopted an insignia in 1911 that includes the Berghaus projection. The lyrics at the end of Chapter 9 are a friendly parody of "Those Were The Days," the theme song of the 1970s television comedy "All in the Family"; my apologies to Lee Adams and Charles Strouse.

"With A Little Help From My Friends" is the title of a song sung by the Beatles, the dominant rock group of the 1960s. The song appeared on the *Sergeant Pepper's* album. "Twist And Shout" was written by Phil Medley and Bert Russell, recorded by the Isley Brothers in 1960, and immortalized by the Beatles on their 1963 albums *Please Please Me* (United Kingdom) and *Introducing the Beatles* (United States). The title for Chapter 17 is a take-off on "As the World Turns," the title of a popular television soap opera that premiered in 1956 and was still being televised in 2001.

After a long illness, Anton Hanegraaf died on September 28, 2001. The Dutch sculptor Rinus Roelofs (2001) has embarked on a project to animate many of Anton's models, which thus may be available in the future on the web.

Using my design, wood craftsman Wayne Daniel produced two wonderfully precise models out of beautiful hardwoods. One model is of Figure 22.7, with the three right triangular pieces out of ipe wood and the remaining four pieces out of peroba rosa wood. The other model is of Figure 1.1. I have demonstrated these models in talks at several conferences.

Wayne also built a wonderfully accurate model of the cyclicly twist-hinged dissection of a hexagram to a hexagon (Figures 22.46 and 22.47). It worked perfectly, transforming from a hexagram to a hexagon and back, except that it could not achieve the intermediate configuration shown in Figure 22.47! It worked all right if I did not hold the trapezoid-centered set of five pieces rigidly flat, allowing them instead to rotate relative to each other. Or I could hold the set of five pieces rigidly flat as long as I did not twist the set of pieces in the same inward (or, alternatively, outward) direction as both large pieces. But the symmetric – and supposedly mathematically correct – motion suggested by Figure 22.47 would always jam at some point. After some thought, I realized that the problem was not in the math but rather in building a three-dimensional model of a two-dimensional object. The symmetric motion founders on the reef of the third dimension: pieces that have thickness (in particular, the triangles marked by asterisks and stars) jam against the pieces that they are hinged with. How appropriate that this dissection, so rich with lovely properties, somewhat cantankerously illustrates one final point.

Cromwell (1997) gives a detailed discussion of the rigidity and flexibility of polyhedra. I had considered writing "Turnabout" sections on foldable tessellations, on rigidity in three dimensions, on linkages, and on hinged polyhedral faces. Readers may enjoy exploring further references by Ian Stewart (*SciAm,* 1999a), Kempe (1877), Graver, Servatius, and Servatius (1993), Hopcroft, Joseph, and Whitesides (1984, 1985), Kantabutra and Kosaraju (1986), van Kreveld, Snoeyink, and Whitesides (1996), Biedl and colleagues (2001), Connelly, Demaine, and Rote (2000), Streinu (2000), and Stuart (1963). Be forewarned that a number of these references are research-oriented.

Erik Demaine asked if there are any 3-way swing-hinged dissections (aside from the hinged dissections of polyregulars). The closeness between the Greek Cross and the square makes several fairly reasonable. The Greek Cross and the square combine with the dodecagon in 12 pieces, with the hexagon in 12 pieces, with the triangle in 14 pieces, and with the Cross of Lorraine in 15 pieces. These 3-way hinged dissections derive from Figures 10.17, 11.44, 11.40, and 10.15 (respectively). Gavin Theobald noticed that you can obtain an 8-piece 3-way swing-hinged dissection of a triangle, square, and hexagram by combining Figures 1.1 and 15.13.

I am maintaining a web-page about the book where I post new developments and items of interest. The URL is:

⟨http://www.cs.purdue.edu/homes/gnf/book2.html⟩

I would like to hear of any additional historical or biographical information, and of any new or improved dissections. My network address is:

gnf@cs.purdue.edu

Bibliography

Abu'l-Wafā' al-Būzjānī. *Kitāb fīmā yahtāju al-sāni' min a' māl al-handasa* (*On the Geometric Constructions Necessary for the Artisan*). Mashdad: Imam Riza 37, copied in the late 10th or the early 11th century. Persian manuscript.

Akiyama, Jin and Gisaku Nakamura (2000a). Congruent Dudeney dissection of polygons. Research Institute of Educational Development, Tokai University.

Akiyama, Jin and Gisaku Nakamura (2000b). Dudeney dissection of polygons. In J. Akiyama, M. Kano, and M. Urabe (Eds.), *Discrete and Computational Geometry, Japanese Conference, JCDCG'98* (Lecture Notes in Computer Science, vol. 1763, pp. 14–29). Berlin: Springer-Verlag.

Akiyama, Jin and Gisaku Nakamura (2000c). Transformable solids exhibition. 32-page color catalog.

Allman, George Johnston (1889). *Greek Geometry from Thales to Euclid*. Dublin: Hodges, Figgis & Co.

Anonymous. *Fī tadākhul al-ashkāl al-mutashābiha aw al-mutawāfiqa* (*Interlocks of Similar or Complementary Figures*). Paris: Bibliothèque Nationale, ancien fonds. Persan 169, ff. 180r–199v.

Appel, Kenneth and Wolfgang Haken (1977). Every planar map is four colorable. *Illinois Journal of Mathematics 21*, 429–567.

Banks, Robert B. (1999). *Slicing Pizzas, Racing Turtles, and Further Adventures in Applied Mathematics*. Princeton, NJ: Princeton University Press.

Bauer, Jean (1999). Polygons' combinations. Powerpoint slide presentation.

Benson, Donald C. (1999). *The Moment of Truth: Mathematical Epiphanies*. Oxford University Press.

Berloquin, Pierre (*Le Monde*). En toute logique. Le Monde. Column in the fortnightly section: des Sciences et des Techniques. (1974a): March 27, p. 18; (1974b): April 10, p. 15; (1974c): May 8, p. 22; (1974d): May 22, p. 20; (1975a): April 23, p. 20.

Biedl, T., E. Demaine, M. Demaine, S. Lazard, A. Lubiw, J. O'Rourke, M. Overmars, S. Robbins, I. Streinu, G. Toussaint, and S. Whitesides (2001). Locked and unlocked polygonal chains in three dimensions. *Discrete & Computational Geometry 26*, 269–81.

Bolyai, Farkas (1832). *Tentamen juventutem*. Maros Vasarhelyini: Typis Collegii Reformatorum per Josephum et Simeonem Kali.

Böttcher, J. E. (1921). Beweis des Tsabît für den pythagoreischen Lehrsatz. *Zeitschrift für mathematischen und naturwissenschaftlichen Unterricht 52*, 153–60.

Bradley, H. C. (1921). Problem 2799. *American Mathematical Monthly 28*, 186–7.

Bradley, H. C. (1930). Problem 3048. *American Mathematical Monthly 37*, 158–9.

Bricard, R. (1897). Mèmoire sur la théorie de l'octaèdre articulé. *J. de Math., Pures et Appliquées 3*, 113–48.

Brodie, B. (1884). Superposition. *Knowledge* 5(135), 399.

Brodie, Robert (1891). Professor Kelland's problem on superposition. *Transactions of Royal Society of Edinburgh 36, part II*(12), 307-11 + plates 1 & 2. See Fig. 12, pl. 2.

Bruyr, Donald L. (1963). *Geometrical Models and Demonstrations*. Portland, ME: J. Weston Walch.

Busschop, Paul (1876). Problèmes de géométrie. *Nouvelle Correspondance Mathématique 2*, 83-4.

Cantor, Moritz (1907). *Vorlesungen über Geschichte der Mathematik*, 3rd ed., vol. 1. Stuttgart: Teubner. Reprinted 1965.

Carpitella, Duilio (1996). Geometric equidecomposition of figures and solids. Unpublished manuscript.

Catalan, Eugène (1873). *Géométrie Élémentaire*, 5th ed. Paris: Dunod. See p. 194.

Cauchy, A. L. (1813). Sur les polygones et les polyèdres. *Second Mémoire, I. Ecole Polytechnique 9*, 87-98.

Cheney, Wm. Fitch (1933). Problem E4. *American Mathematical Monthly 40*, 113-14.

Collison, David M. (1979). Rational geometric dissections of convex polygons. *Journal of Recreational Mathematics 12*(2), 95-103.

Connelly, Robert (1978a). Conjectures and open problems in rigidity. In *Proceedings of the International Congress of Mathematicians*, Helsinki, pp. 407-14.

Connelly, Robert (1978b). A counter-example to the rigidity conjecture for polyhedra. *Publications Math. de l'Institut des Hautes Études Scientifiques 47*, 333-8.

Connelly, Robert (1979). The rigidity of polyhedral surfaces. *Mathematics Magazine 52*, 275-83.

Connelly, Robert, Erik D. Demaine, and Günter Rote (2000). Straightening polygonal arcs and convexifying polygonal cycles. In *Proceedings of the 41st IEEE Symposium on the Foundations of Computer Science*, pp. 432-42.

Connelly, R., I. Sabitov, and A. Walz (1997). The bellows conjecture. *Contributions to Algebra and Geometry 38*, 1-10.

Cooper, L. A. and R. N. Shepard (1973). Chronometric studies of the rotation of mental images. In W. G. Chase (Ed.), *Visual Information Processing*. New York: Academic Press.

Cromwell, Peter (1997). *Polyhedra*. Cambridge University Press.

Crowe, D. W. and I. J. Schoenberg (1984). On the equidecomposability of a regular triangle and a square of equal areas. *Mitteilungen aus dem mathematische Seminar Giessen 164*, 59-63. In the Coxeter-Festschrift, part II.

Cundy, H. M. and C. D. Langford (1960). On the dissection of a regular polygon into *n* equal and similar parts. *Mathematical Gazette 44*, 46.

Cundy, H. Martyn and A. P. Rollett (1952). *Mathematical Models*. Oxford University Press.

de Coatpont (1877). Sur un problème de M. Busschop. *Nouvelle Correspondance Mathématique*, 116-17.

Dehn, M. (1900). Über den Rauminhalt. *Nachrichten von der Gesellschaft der Wissenschaften zu Göttingen, Mathematisch-Physikalische Klasse*, 345-54. Subsequently in *Mathematische Annalen 55* (1902), 465-78.

Demaine, Erik D., Martin L. Demaine, David Eppstein, and Erich Friedman (1999). Hinged dissection of polyominoes and polyiamonds. In *Proceedings of the 11th Canadian Conference on Computational Geometry*, Vancouver. Electronic proceedings at ⟨http://www.cs.ubc.ca/conferences/CCCG/elec_proc/elecproc.html⟩.

Dickson, L. E. (1920). *History of the Theory of Numbers*, vol. II: *Diophantine Analysis*. New York: Chelsea.

Dudeney, Henry Ernest (1907). *The Canterbury Puzzles and Other Curious Problems*. London: W. Heinemann. Revised edition printed by Dover Publications in 1958.

Dudeney, Henry Ernest (1917). *Amusements in Mathematics.* London: Thomas Nelson & Sons. Revised edition printed by Dover Publications in 1958.

Dudeney, Henry Ernest (1926). *Modern Puzzles and How to Solve Them.* London: C. Arthur Pearson.

Dudeney, Henry Ernest (1931). *Puzzles and Curious Problems.* London: Thomas Nelson & Sons.

Dudeney, Henry E. (*Dispatch*). Puzzles and prizes. Column in *Weekly Dispatch,* April 19, 1896–December 26, 1903. (1900a): August 26; (1902a): November 16.

Dudeney, Henry E. (*London*). The Canterbury puzzles. Series in *London Magazine,* January 1902–November 1903. (1903a): November, p. 443.

Dudeney, Henry E. (*Strand*). Perplexities. Monthly puzzle column in *The Strand Magazine* from 1910 to 1930. (1911a): vol. 42, p. 108; (1920a): vol. 60, pp. 368, 452; (1923a): vol. 65, p. 405; (1926a): vol. 71, p. 416; (1926b): vol. 71, p. 522; (1926c): vol. 72, p. 316, (1927a): vol. 73, pp. 305, 420.

Dudeney, Henry E. (*Tit-Bits*). Weekly puzzle column in *Tit-Bits,* September 25, 1897, continuing into 1898. (1897a): November 13, p. 119.

Dudeney, Mrs. Henry (1998). *A Lewes Diary, 1916-1944.* Lewes: Tartarus Press. Edited by Diana Crook.

Elliott, C. S. (1982). Some new geometric dissections. *Journal of Recreational Mathematics 15*(1), 19-27.

Elliott, C. S. (1985). Some more geometric dissections. *Journal of Recreational Mathematics 18*(1), 9-16.

Elliott, C. S. (1995). An extension to polygram shapes and some new dissections. *Journal of Recreational Mathematics 27*(4), 277-84.

Elser, Veit and Michael Goldberg (1977). Solution to problem 969, cube covering. *Mathematics Magazine 50,* 168-9.

Eppstein, David (2001). Hinged kite mirror dissection. ACM Computing Research Repository, cs.CG/0106032, ⟨http://arxiv.org/abs/cs.CG/0106032⟩.

Esser, William L. III (1985). Jewelry and the like adapted to define a plurality of objects or shapes. U.S. Patent 4,542,631, filed 1983.

Eves, Howard (1963). *A Survey of Geometry,* vol. I. Boston: Allyn & Bacon.

Eves, Howard W. (1972). *Mathematical Circles Squared.* Boston: Prindle, Weber & Schmidt.

Fourrey, E. (1907). *Curiosités Géométriques.* Paris: Vuibert et Nony.

Frederickson, Greg N. (1972a). Appendix H: Eight years after. In Harry Lindren, *Recreational Problems in Geometric Dissections and How to Solve Them.* New York: Dover Publications.

Frederickson, Greg N. (1972b). Assemblies of twelve-pointed stars. *Journal of Recreational Mathematics 5*(2), 128-32.

Frederickson, Greg N. (1972c). Polygon assemblies. *Journal of Recreational Mathematics 5*(4), 255-60.

Frederickson, Greg N. (1972d). Several star dissections. *Journal of Recreational Mathematics 5*(1), 22-6.

Frederickson, Greg N. (1974). More geometric dissections. *Journal of Recreational Mathematics 7*(3), 206-12.

Frederickson, Greg N. (1997). *Dissections: Plane & Fancy.* Cambridge University Press.

Friedman, Erich (2000). Problem of the month (January 2000). ⟨http://www.stetson.edu/~efriedma/mathmagic/0100.html⟩.

Gantes, C., A. Giakoumakis, and P. Vousvounis (1997). Symbolic manipulation as a tool for a design of deployable domes. *Computers & Structures 64,* 865-78.

Gardner, Martin (1958). Problem E1309. *American Mathematical Monthly 65*(3), 205.

Gardner, Martin (1961). *The 2nd Scientific American Book of Mathematical Puzzles & Diversions*. New York: Simon & Schuster.

Gardner, Martin (1969). *The Unexpected Hanging and Other Mathematical Diversions*. University of Chicago Press. Updated with a new afterward and an expanded bibliography, 1991.

Gardner, Martin (1997). *The Last Recreations: Hydras, Eggs, and Other Mathematical Mystifications*. New York: Copernicus, Springer-Verlag.

Gardner, Martin (*SciAm*). Mathematical games. Monthly column in *Scientific American*. (1963a): November, p. 144; (1963b): December, pp. 144, 146; (1964a): February, p. 128; (1980a): February, pp. 14-19.

Gerwien, P. (1833). Zerschneidung jeder beliebigen Anzahl von gleichen geradlinigen Figuren in dieselben Stücke. *Journal für die reine und angewandte Mathematik (Crelle's Journal) 10*, 228-34 and Taf. III.

Gluck, H. (1975). Almost all simply connected closed surfaces are rigid. In *Geometric Topology* (Lecture Notes in Mathematics, vol. 438, pp. 225-39). Berlin: Springer-Verlag.

Goldberg, Michael (1940). Solution to problem E400. *American Mathematical Monthly 47*, 490-1.

Goldberg, Michael (1952). Problem E972: Six piece dissection of a pentagon into a triangle. *American Mathematical Monthly 59*, 106-7.

Goldberg, Michael (1966). A duplication of the cube by dissection and a hinged linkage. *Mathematical Gazette 50*, 304-5.

Goldberg, Michael (1978). Unstable polyhedral structures. *Mathematics Magazine 51*, 165-70.

Golomb, Solomon W. (1994). *Polyominoes: Puzzles, Patterns, Problems, and Packings*, 2nd ed. Princeton, NJ: Princeton University Press.

Graver, Jack, Brigitte Servatius, and Herman Servatius (1993). *Combinatorial Rigidity*. Providence, RI: American Mathematical Society.

Hadwiger, H. and P. Glur (1951). Zerlegungsgleichheit ebener polygone. *Elemente der Mathematik 6*, 97-106.

Hanegraaf, Anton (1989). The Delian altar dissection. Elst, the Netherlands. First booklet in his projected series *Polyhedral Dissections*.

Hart, Harry (1877). Geometrical dissections and transpositions. *Messenger of Mathematics 6*, 150-1.

Heawood, P. J. (1890). Map colour theorems. *Quarterly Journal of Pure and Applied Mathematics 24*, 332-8.

Hilbert, David (1900). Mathematische probleme. *Nachrichten von der Gesellschaft der Wissenschaften zu Göttingen, Mathematisch-Physikalische Klasse*. Subsequently in *Bulletin of the American Mathematical Society* 8 (1901-1902), 437-79.

Hoffman, Frederick J. (1945). *Freudianism and the Literary Mind*. New York: Grove Press. See his comments on Freud's *Interpretation of Dreams* on pp. 38-40.

Hopcroft, John, Deborah Joseph, and Sue Whitesides (1984). Movement problems for 2-dimensional linkages. *SIAM Journal on Computing 13*, 610-29.

Hopcroft, John, Deborah Joseph, and Sue Whitesides (1985). On the movement of robot arms in 2-dimensional bounded regions. *SIAM Journal on Computing 14*, 315-33.

Hussey, Maurice (1968). *Chaucer's World: A Pictorial Companion*. Cambridge University Press.

Jäger, G. (1867). Der Nordpol, ein thiergeographisches Centrum. *Mittheilungen aus Justus Perthes' geographischer Anstalt über wichtige neue Erforschungen auf dem Gesamtgebiete der Geographie, von Dr. A. Petermann, Ergänzungsband 4*(16), 67-70 and Tafel 3.

Kantabutra, V. and S. R. Kosaraju (1986). New algorithms for multilink robot arms. *Journal of Computer and Systems Sciences 32*, 136-53.

Kelland, Philip (1855). On superposition. *Transactions of the Royal Society of Edinburgh 21*, 271-3 and plate V.

Kelland, Philip (1864). On superposition. Part II. *Transactions of the Royal Society of Edinburgh 33*, 471-3 and plate XX.

Kempe, A. B. (1877). *How to Draw a Straight Line; A Lecture on Linkages.* London: Macmillan.

King, Francis and Bruce Steele (1969). *Chaucer: The Prologue and Three Tales.* London: John Murray.

Klee, Victor and Stan Wagon (1991). *Old and New Unsolved Problems in Plane Geometry and Number Theory* (Dolciani Mathematical Expositions). Washington, DC: Mathematical Association of America.

Kramer, Glenn A. (1992). *Solving Geometric Constraint Systems.* Cambridge, MA: MIT Press.

Kürschák, Josef (1899). Über das regelmässige Zwölfeck. *Mathematische und naturwissenschaftliche Berichte aus Ungarn 15*, 196-7.

Laczkovich, Miklós (1990). Equidecomposability and discrepancy; a solution of Tarski's circle-squaring problem. *Journal für die reine und angewandte Mathematik 404*, 77-117.

Langford, C. Dudley (1956). To pentasect a pentagon. *Mathematical Gazette 40*, 218.

Langford, C. Dudley (1967a). On dissecting the dodecagon. *Mathematical Gazette 51*, 141-2.

Langford, C. Dudley (1967b). Polygon dissections. *Mathematical Gazette 51*, 139-41.

Lemon, Don (1890). *The Illustrated Book of Puzzles.* London: Saxon.

Leonardo da Vinci (1973). *Il Codice Atlantico.* Firenze: Giunti-Barbera. Facsimile reproduction of the *Codex Atlanticus,* in twelve volumes. (5a): vol. 5, p. 368; (5b): vol. 5, p. 442; (6a): vol. 6, p. 453; (6b): vol. 6, p. 463; (6c): vol. 6, p. 540; (12a): vol. 12, p. 1081.

Lindgren, H. (1951). Geometric dissections. *Australian Mathematics Teacher 7*, 7-10.

Lindgren, H. (1953). Geometric dissections. *Australian Mathematics Teacher 9*, 17-21, 64.

Lindgren, H. (1956). Problem E1210: A dissection of a pair of equilateral triangles: Solution. *American Mathematical Monthly 63*, 667-8.

Lindgren, H. (1958). Problem E1309: Dissection of a regular pentagram into a square. *American Mathematical Monthly 65*, 710-11.

Lindgren, H. (1960). A quadrilateral dissection. *Australian Mathematics Teacher 16*, 64-5.

Lindgren, H. (1961). Going one better in geometric dissections. *Mathematical Gazette 45*, 94-7.

Lindgren, Harry (1962). Three Latin-cross dissections. *Recreational Mathematics Magazine* (8), 18-19. April.

Lindgren, H. (1964a). Dissections for schools. *Australian Mathematics Teacher 20*(3), 52-4. The figures are in the supplement to the issue, pp. i, ii.

Lindgren, Harry (1964b). *Geometric Dissections.* Princeton, NJ: Van Nostrand.

Lindgren, Harry (1970). A dissection problem by Sam Loyd. *Journal of Recreational Mathematics 3*(1), 54-5.

Lindgren, Harry (1972). *Recreational Problems in Geometric Dissections, and How to Solve Them.* New York: Dover Publications.

Loeb, Arthur L. (1976). *Space Structures: Their Harmony and Counterpoint.* Reading, MA: Addison-Wesley.

Loyd, Sam (*Eliz. J.*). Puzzle column in *Elizabeth Journal.* (1908a): October 29.

Loyd, Sam (*Home*). Sam Loyd's own puzzle page. Monthly puzzle column in *Woman's Home Companion,* 1903-1911. (1908a): November, p. 51.

Loyd, Sam (*Inquirer*). Mental gymnastics. Puzzle column in Sunday edition of *Philadelphia Inquirer,* October 23, 1898-1901. (1901a): August 11.

Loyd, Sam (*Tit-Bits*). Weekly puzzle column in *Tit-Bits,* starting in October 3, 1896, and continuing into 1897. Dudeney, under the pseudonym of "Sphinx," wrote commentary and handled the awarding of prize money. Later, in 1897, Dudeney took over the column. (1897a): April 3, p. 3; (1897b): April 24, p. 59; (1897c): July 17, p. 291.

Lucas, Édouard (1883). *Récréations Mathématiques,* vol. 2. Paris: Gauthier-Villars. Second of four volumes. Second edition (1893) reprinted by Blanchard in 1960; see pp. 151-2 in vol. 2 of this edition.

Lurker, Ernst (1984). Heart pill. Model in nickel-plated aluminum, limited edition of 80 produced by Bayer, in Germany.

Lurker, Ernst (1985). *Play Art and Creativity.* Bayer. 48 pp.

Macaulay, W. H. (1914). The dissection of rectilineal figures. *Mathematical Gazette 7,* 381-8.

Macaulay, W. H. (1915). The dissection of rectilineal figures. *Mathematical Gazette 8,* 72-6, 109-15.

Macaulay, W. H. (1919). The dissection of rectilineal figures (continued). *Messenger of Mathematics 49,* 111-21.

Macaulay, W. H. (1922). The dissection of rectilineal figures (continued). *Messenger of Mathematics 52,* 53-6.

MacMahon, Percy A. (1922). Pythagoras's theorem as a repeating pattern. *Nature 109,* 479.

Mahlo, Paul (1908). *Topologische Untersuchungen über Zerlegung in ebene und sphaerische Polygone.* Halle, Germany: C. A. Kaemmerer. Ph.D. dissertation for the Vereinigte Friedrichs-Universität in Halle-Wittenberg. See pp. 13-14 and Fig. 7.

Maksimov, I. G. (1995). Polyhedra with bendings and Riemann surfaces. *Russian Mathematical Surveys 50*(4), 821-3. Translated from *Uspekhi Matematicheskikh Nauk 50*(4), 163-4.

McCabe, James Edward (1972). *Leonardo da Vinci's De Ludo Geometrico.* Ph.D. thesis, University of California, Los Angeles.

McKenna, Katharine (1988). Geometric toy. U.S. Patent 4,722,712, filed 1985.

Mott-Smith, Geoffrey (1946). *Mathematical Puzzles for Beginners and Enthusiasts.* Philadelphia: Blakiston. Reprinted by Dover Publications, New York, 1954.

Mulder, Henk (1977). Transformatie: Binnenste-buiten. *Pythagoras 17,* 1-5.

Needham, Joseph (1959). *Science and Civilisation in China,* vol. 3: *Mathematics and the Sciences of the Heavens and the Earth.* Cambridge University Press. See Fig. 52 on p. 29 and discussion on p. 27.

Özdural, Alpay (2000). Mathematics and arts: Connections between theory and practice in the medieval Islamic world. *Historia Mathematica 27,* 171-201.

Paterson, David (1988). Two dissections in 3-D. *Journal of Recreational Mathematics 20*(4), 257-70.

Paterson, David (1989). T-dissections of hexagons and triangles. *Journal of Recreational Mathematics 21*(4), 278-91.

Pattison, William D. (1960). The star of the AAG. *Professional Geographer 12*(5), 18-19.

Perigal, Henry (1873). On geometric dissections and transformations. *Messenger of Mathematics 2,* 103-5.

Perigal, Henry (1875). Geometrical dissections and transformations, no. II. *Messenger of Mathematics 4,* 103-4.

Perigal, Henry (1891). *Graphic Demonstrations of Geometric Problems.* London: Bowles & Sons. On cover: "Association for the Improvement of Geometrical Teaching. Geometric Dissections and Transpositions." (The association was later renamed The Mathematical Association.)

Pinker, Steven (1997). *How the Mind Works.* New York: Norton.

Reid, Robert (1987). Disecciones geometricas. *Umbral* (2), 59-65. Published in Lima, Peru, by Asociacion Civil Antares. The author's name as listed in the article is Robert Reid Dalmau, conforming to Spanish custom, but is listed here in the form that Robert prefers.

Resch, Ronald D. (1965). Geometrical device having articulated relatively movable sections. U.S. Patent 3,201,894, filed 1963.

Ringel, Gerhard (1959). *Färbungsprobleme auf Flächen und Graphen*. Berlin: Deutscher Verlag der Wissenschaften.

Roelofs, Rinus (2001). Anton Hanegraaf's dissection models (working title). May be made available on the web; see Roelofs's website ⟨http://www.itebo.nl/rinus/Index.html⟩. Animations of Anton Hanegraaf's models.

Rosenbaum, Joseph (1947). Problem E721: A dodecagon dissection puzzle. *American Mathematical Monthly 54*, 44.

Rubik, Erno (1983). Toy with turnable elements for forming geometric shapes. U.S. Patent 4,392,323, filed 1981; filed for a Hungarian patent in 1980.

Sah, Chih-han (1979). *Hilbert's Third Problem: Scissors Congruence* (Research Notes in Mathematics, vol. 33). San Francisco: Pitman.

Sayili, Aydin (1960). Thâbit ibn Qurra's generalization of the Pythagorean theorem. *Isis 51*, 35–7.

Schmerl, James (1973). Problem #240. A Pythagorean dissection. *Journal of Recreational Mathematics 6*(4), 315–16. See also vol. *7*(2), 153 (1974).

Schoenberg, I. J. (1982). *Mathematical Time Exposures*. Washington, DC: Mathematical Association of America.

Shen Kangshen, John N. Crossley, and Anthony W.-C. Lun (1999). *The Nine Chapters on the Mathematical Art: Companion and Commentary*. Oxford University Press & Beijing: China Press. See pp. 123–5.

Slothouber, Jan (1973). Flexicubes – Reversible cubic shapes. *Journal of Recreational Mathematics 6*, 39–46.

Snyder, John P. (1993). *Flattening the Earth: Two Thousand Years of Map Projections*. University of Chicago Press.

Steffen, Klaus (1978). A symmetric flexible Connelly sphere with only nine vertices. Unpublished letter, Institut des Hautes Études Scientifiques.

Steinhaus, Hugo (1960). *Mathematical Snapshots*, 2nd ed. New York: Oxford University Press. See p. 11.

Steinhaus, Hugo (1969). *Mathematical Snapshots*, 3rd ed. New York: Oxford University Press.

Steinitz, E. and H. Rademacher (1934). *Vorlesungen über die Theorie der Polyedern*. Berlin: Springer-Verlag.

Stewart, Ian (1987). *The Problems of Mathematics*. Oxford University Press.

Stewart, I. (*SciAm*). Mathematical recreations. Column in *Scientific American*. (1999a): "Origami Tessellations," February, pp. 100–1.

Streinu, I. (2000). A combinatorial approach to planar non-colliding robot arm motion planning. In *Proceedings of the 41st IEEE Symposium on the Foundations of Computer Science*, pp. 843–53.

Stuart, Duncan (1963). Polyhedral and mosaic transformations. *Student Publications of the School of Design, North Carolina State University 12*(1), 2–28. The finger movies make up the left-hand portion of the issue.

Sturm, Johann Cristophorus (1700). *Mathesis Enumerata: Or, the Elements of the Mathematicks*. London: Robert Knaplock. Translation (by J. Rogers?) of 1695 work *Mathesis Enumerata*. See pp. 20–1 and Fig. 29.

Tarr, Micheal J. (1995). Rotating objects to recognize them: A case study on the role of viewpoint dependency in the recognition of three-dimensional shapes. *Psychonomic Bulletin and Review 2*, 55–82.

Tarr, Micheal J. and Steven Pinker (1989). Mental rotation and orientation-dependence in shape recognition. *Cognitive Psychology 21*, 233–82.

Taylor, H. M. (1905). On some geometrical dissections. *Messenger of Mathematics 35*, 81–101.

279

Taylor, Henry Martin (1909). *Mathematical Questions and Solutions from "The Educational Times"* 16, 81-2. Second series.

Tilson, Philip Graham (1978). New dissections of pentagon and pentagram. *Journal of Recreational Mathematics 11*(2), 108-11.

Tjebbes, T. (1969). ABT Abstract, nu concreet. *ABT-mededelingen 1*(2), 1-6. Photoreport of an exposition in Arnheim. See photograph on lower left of p. 6.

Van den Broeck, Luc (1997). Twee problemen scharnierend rond een puzzel. *Wiskunde & Onderwijs 23*(92), 309-17.

van Kreveld, Mark, Jack Snoeyink, and Sue Whitesides (1996). Folding rulers inside triangles. *Discrete Computational Geometry 15*, 265-85.

Varsady, Alfred (1986). The dissection of sets of polygons. *Journal of Recreational Mathematics 18*(4), 256-68.

Varsady, Alfred (1989). Some new dissections. *Journal of Recreational Mathematics 21*(3), 203-9.

Verrill, Helena (1998). Origami tessellations. In R. Sarhangi (Ed.), *Proceedings of the First Annual Conference, Bridges: Mathematical Connections in Art, Music, and Science*, pp. 55-68.

Wagon, Stan (1985). *The Banach-Tarski Paradox* (Encyclopedia of Mathematics and its Applications, vol. 24). Cambridge University Press.

Wallace, William (Ed.) (1831). *Elements of Geometry*, 8th ed. Edinburgh: Bell & Bradfute. First six books of Euclid, with a supplement by John Playfair.

Wasfy, Tamer M. and Ahmed K. Noor (1998). Application of fuzzy sets to transient analysis of space structures. *Finite Elements in Analysis and Design 29*, 153-71.

Weisstein, Eric W. (1998). *CRC Concise Encyclopedia of Mathematics*. Boca Raton, FL: CRC Press.

Wells, David (1975). Figures: On gems and generalisations. *Games & Puzzles* (38), 40. June. Author not identified, but attributed to David Wells in (Wells 1991).

Wells, David (1988). *Hidden Connections, Double Meanings*. Cambridge University Press.

Wells, David (1991). *The Penguin Dictionary of Curious and Interesting Geometry*. London: Penguin Books.

Wheeler, A. H. (1935). Problem E4. *American Mathematical Monthly 42*, 509-10.

Wills, Herbert III (1985). *Leonardo's Dessert, No Pi*. Reston, VA: National Council of Teachers of Mathematics.

Winkler, A. (1929). Ein Modell, das alle mögliche Fälle des Lehrsatzes des Pythagoras veranschaulicht. *Zeitschrift für mathematischen und naturwissenschaftlichen Unterricht 60*, 49-52.

Woepcke, F. (1855). Analyse et extrait d'un recueil de constructions géométriques par Aboûl Wafâ. *Journal Asiatique V*, 318-59.

Wunderlich, Walter (1965). Starre, kippende, wackelige und bewegliche Achtfläche. *Elemente der Mathematik 20*, 25-32.

Yates, Robert C. (1940). Addition by dissection. *School Science and Mathematics 40*, 801-7.

Index of Dissections

Dissections are ordered by the following conventions.

1. Two-dimensional dissections precede three-dimensional dissections.

2. The two-dimensional figures are ordered with those represented by letters (in alphabetical order) following those represented by $\{p\}$ or $\{p/q\}$ (in lexicographic order on (p,q)).

3. A dissection is listed under the figure it involves that comes latest in the list.

4. For a figure such as $\{r\}$, dissections of it involving $\{p\}$ or $\{p/q\}$ with $p < r$ come first. Within dissections only involving $\{r\}$, special relationships come first (ordered lexicographically), then general relationships (ordered lexicographically), then a $\{r\}$s to b $\{r\}$s (ordered lexicographically on (a,b)).

281

282

General Index